CO$_2$ Sequestration
by Ex-Situ Mineral Carbonation

CO$_2$ Sequestration
by Ex-Situ Mineral Carbonation

Editors

Aimaro Sanna
M. Mercedes Maroto-Valer
Heriot-Watt University, UK

World Scientific

NEW JERSEY · LONDON · SINGAPORE · BEIJING · SHANGHAI · HONG KONG · TAIPEI · CHENNAI · TOKYO

Published by

World Scientific Publishing Europe Ltd.

57 Shelton Street, Covent Garden, London WC2H 9HE

Head office: 5 Toh Tuck Link, Singapore 596224

USA office: 27 Warren Street, Suite 401-402, Hackensack, NJ 07601

Library of Congress Cataloging-in-Publication Data
Names: Sanna, Aimaro, editor. | Maroto-Valer, M. Mercedes, editor.
Title: CO_2 sequestration by ex-situ mineral carbonation / [compiled by] Aimaro Sanna
    (Heriot-Watt University, UK), Mercedes Maroto-Valer (Heriot-Watt University, UK).
Description: [Hackensack?] New Jersey : World Scientific, [2017] |
    Includes bibliographical references.
Identifiers: LCCN 2016026158 | ISBN 9781786341594 (hc : alk. paper)
Subjects: LCSH: Carbon dioxide mitigation. | Carbon sequestration. |
    Carbonate minerals. | Silicate minerals. | Artificial minerals.
Classification: LCC TD885.5.C3 C6478 2017 | DDC 628.5/32--dc23
LC record available at https://lccn.loc.gov/2016026158

British Library Cataloguing-in-Publication Data
A catalogue record for this book is available from the British Library.

Desk Editors: Chandrima Maitra/Mary Simpson

Typeset by Stallion Press
Email: enquiries@stallionpress.com

Preface

The global atmospheric CO_2 level surpassed 400 ppm in March 2015 for the first time on records, according to the US National Oceanic and Atmospheric Administration (NOAA), which means that we have added 120 ppm CO_2 to the atmosphere since pre-industrial times, mainly burning fossil fuels. Moreover, since 1992 (when the developed countries signed a binding agreement to reduce CO_2 emissions called Kyoto Protocol), the global economic scenario and consecutively the global CO_2 emissions share have drastically changed, requiring a direct involvement of the new emerging economic powers such as China, India, Brazil, etc., which are now major CO_2 emitters in the effort to tackle down the CO_2 level to maintain the atmospheric temperature rise within 2°C.

A wake-up call arrived in 2015, thanks to the latest United Nations Conference on Climate Change (COP21) held in Paris (November 30–December 12), where 195 countries adopted the first universal climate agreement. Intended Nationally Determined Contributions (INDCs) submitted for the Paris agreement cover 186 countries, representing about 90% of global emissions, compared to about 15% achieved by the Kyoto Protocol in 1992.

The IEA World Energy Outlook Special Briefing for COP21 shows that even if INDCs pledges will be honoured, worldwide, they will fall short of the major course correction required to achieve the agreed climate goal, resulting in average global temperature increase of around 2.7°C by 2100. This would require zero-virtual emissions from the world major CO_2 emitters post-2030. Moreover, the 2015 IEA World Energy Outlook — Special Report indicates that without doubt fossil fuels will still represent the primary source of energy up to 2030, even in the INDC scenario.

Nevertheless, the numbers above clearly suggest that we must act now to change how we use and generate energy and how we use our land if we want to limit the rise in global mean temperature to $2°C$. Despite the fact that the Paris Agreement represents a key step towards tackling climate change, the problem has not been solved and it will require not only that the IMDCs are met in full, but also a series of additional actions in the years to come.

The issues surrounding the high levels of anthropogenic CO_2 in the atmosphere are challenging the scientific community to find a way to prevent, and at the very least, further significant increases. It is generally accepted that to address this issue, a portfolio of different and complementary technologies such as renewables, changes in energy use and carbon capture and storage (CCS) must be employed.

In this context, mineral carbonation (MC) is emerging as a potential technology to sequester CO_2. In MC, CO_2 emitted from industrial and power plants chemically reacts with calcium- and/or magnesium-containing materials to form stable carbonates. Two advantages make MC unique amongst the different storage approaches: the abundance of metal oxide bearing materials, particularly natural silicates, and the permanence of storage of CO_2 in a stable solid form. MC, which has been under development for more than 20 years, has the potential to sequester the totality of the CO_2 emitted assuming burning of all the global fossil fuel reserves. Initially, MC was considered only as an alternative to underground storage of separated CO_2, but because it produces stable carbonates that cannot release back in the atmosphere, the binded CO_2 in ambient conditions, it has become increasingly attractive and currently is being investigated at the commercial scale (e.g. Skyonic process). The renewed interest in MC is related to the possibility to avoid CO_2 capture, which is required for geological storage.

The conception of this book started with the observation that despite the description of mineral carbonation in a number of book chapters, a dedicated book on this topic is not available. Moreover, since the major developments on MC focus on engineered process on the ground (*ex-situ* MC), we decided to focus this book on this type of technology. This book brings together leading experts working in MC to provide an introduction and a critical assessment of progress to date. The book chapters cover the basics of the technology (Chapter 1), resources available worldwide (Chapter 2) as well as *ex-situ* technologies (Chapters 3 and 4) and the potential utilisation of the carbonation products (Chapter 5). In this book, the authors critically evaluate progress to date and identify the key technical and economical obstacles that still need to be addressed. A brief description of the book content is provided below.

Chapter 1: "Mineral Carbonation Technology Overview". This introductory chapter describes the causes of climate change and greenhouse effect, MC in the context of options to fight climate change, the chemistry behind MC and also provides a summary of advantages and challenges, which are treated in detail in the following chapters.

Chapter 2: "Rocks: Cation Donors with Enormous Resources". This chapter describes the range of rock compositions and the progression from (a) natural weathering of these rocks, (b) schemes for accelerated weathering. The chapter contains a description and classification of mafic and ultramafic rocks. The reactivity of minerals indicating the suitability of different types of rocks for MC is treated in this chapter. Also, the global distribution and resources of ultramafic rocks suitable for *ex-situ* MC are discussed here.

Chapter 3: "CO_2 Mineralisation as a Route to Energy-Efficient CO_2 Sequestration or Materials with Market Value". Currently, activation of primarily magnesium silicate-containing rocks using heat, mechanical action or relative mild extraction of magnesium using ammonium salts are the main routes under development. However, although reasonable chemical kinetic rates have been achieved, the energy input/output balance of many suggested CO_2 mineralisation routes still needs to be improved. In those cases, where the process is a net CO_2 producer, this CCS route can only be motivated by the production of marketable materials. Chapter 3 gives an overview of the state-of-the-art of CO_2 mineral sequestration for large-scale CCS/CCSU, focusing on the work done during recent years. For as far data are available, MC is assessed in terms of energy efficiency, solid materials produced and their possible use and water and life cycle assessment.

Chapter 4: "MC Technologies Developed for Waste Materials". This chapter includes an overview of the main characteristics of suitable waste materials and their potential CO_2 uptake. Also, the process chemistry, reaction kinetics, process conditions of the main routes (direct and indirect carbonation) of waste-based MC processes are discussed. The proposed process models and cost estimations as well as the effect of waste remediation are also covered.

Chapter 5: "MC Process Scale and Product Applications". As with any other industrial process, MC processes must satisfy several criteria before they can be implemented on a commercial scale. Among the most important considerations are the economics and pathways for downstream product utilisation. In this chapter, the scale of MC is defined and compared with other similarly sized industries. It is shown that the material flows for MC, although large, are comparable with some industries such as construction and mining. The high- and low-tech applications of the products of MC are also discussed. Large-scale uses for the carbonate/silica mixture include asphalt manufacture, earthwork projects and iron and steel refining. This chapter shows that the markets for all of these applications are suitably large to accommodate the annual production capacity required for meaningful CO_2 sequestration. Combinations and integration of pressure and pH-swing carbonation processes can yield mutual benefits, such that a net profit and CO_2 sequestration is achieved.

Being convinced of the value of mineral carbonation as an approach to fight climate change, we were delighted to collaborate with some of the major experts in our research community on this book. We hope that the readers enjoy reading this book and become familiar with the opportunities and challenges of MC and help us to diffuse and share the content of this book with the general public.

About the Authors

Dr Aimaro Sanna's research group is currently working of the development of low carbon technologies. In 2014, he was awarded a tenure-track fellowship at the School of Engineering and Physical Sciences, Heriot-Watt University. He did is PhD in Environmental Engineering at the University of Nottingham working on the development of biofuels. His scientific interests include biomass conversion into biofuels and biochemicals by thermochemical processes, heterogeneous catalysis, hydro-treating technologies, carbon capture and storage by mineral carbonation and novel materials. Dr Sanna has more than 70 publications in the areas of biofuels and carbon capture and storage. He is Managing Editor for the Elsevier journal *Fuel Processing Technology* and member of the RSC, IChemE, UKCCSRC and CO2Chem.

Prof M. Mercedes Maroto-Valer (FRSE, FIChemE, FRSC, FRSA) is Director of the Centre for Innovation in Carbon Capture and Storage (CICCS) at Heriot-Watt University. She leads a team of 30 researchers developing novel solutions to meet the worldwide strive for cost-effective and environmentally friendly energy, with particular emphasis on clean energy technologies, including carbon dioxide capture, transport, storage and utilization. She has over 400 publications, of which she has been the editor of four books. She holds leading positions in professional societies and editorial boards and has received numerous prestigious international prizes and awards. She has recently received a prestigious European Research Council (ERC) Advanced Award.

Dr Qi Liu is an Associate Professor at the Research Institute of Enhanced Oil Recovery of China University of Petroleum (Beijing). He was Research Associate at the Centre for Innovation in Carbon Capture and Storage (CICCS) of Heriot-Watt University from July 2012 to January 2016. Dr Liu obtained a PhD in Chemical Engineering in 2012 at the University of Nottingham (UK). His research focused on mineral carbonation (MC), experimental and modelling studies on the mineralogical changes and fluid chemistry derived from the injection of CO_2 and co-injection of gas mixtures and geological sequestration of CO_2 in saline aquifers, wellbore integrity and CO_2-EOR.

Dr Mai Uibu (www.etis.ee/Portal/Persons) is currently conducting her research as a Senior Research Scientist at the Laboratory of Inorganic Materials, Tallinn University of Technology. She received her PhD degree in Chemical and Materials Technology in 2008 from Tallinn University of Technology, Estonia. Her research interests include the basics of new utilization processes for inorganic wastes, focusing on CO_2 abatement and sustainable processing of oil shale ashes. She has published more than 80 publications, including 32 referred publications in journals, and over 50 contributions to conferences proceedings.

Regiina Viires (BSc) is currently a master student in Applied Chemistry and Biotechnology at Tallinn University of Technology. She has worked as a Laboratory Assistant at the Laboratory of Inorganic Materials since 2015 and is currently a co-author of a conference paper.

Dr Rein Kuusik (www.etis.ee/Portal/Persons) is one of founders of the Laboratory of Inorganic Materials at Tallinn University of Technology being also a long-term director and scientific leader of the lab. The laboratory carries out broad basic and applied research in the field of chemistry and technology of heterogeneous mineral-organic multicomponent systems, in particular with aim to find out the new utilization are as for Estonian natural resources (oil shale, phosphorites, glauconites, limestones–dolomites, etc.) as well as for inorganic industrial wastes including these ones from oil shale processing industry. He is author or co-author of over 300 publications. He has received state scientific awards during 1975–2010.

Dr Jie Bu is a Senior Scientist at the Institute of Chemical & Engineering Sciences (ICES), Singapore. He received his PhD in Chemical Engineering from Seoul National University of South Korea in 2000 and joined ICES. His current research interests include exploring CO_2 capture and sequestration technology, chemical process economic evaluation and climate change policy study.

Mr Tze Yuen Yeo is a Senior Research Engineer at the Institute of Chemical & Engineering Sciences (ICES). He joined ICES in 2010 after completing his BSc in Chemistry at Nanyang Technological University in Singapore. His current research interests include CO_2 capture and sequestration and separation operations for chemical processes.

Dr Alicja M. Lacinska is a mineralogist/petrologist at the British Geological Survey with several years of experience working on carbon capture and storage by mineralization of ultramafic rocks and serpentinites.

Mike T. Styles is a geologist and Honorary Researcher at the British Geological Survey. He has 40 years experience in the study of mafic and ultramafic rocks from many parts of the world. Over the last 10 years, he has applied this expertise to CCSM. He has worked in teams in collaboration with chemical engineers developing technology and particularly in selecting the most suitable feed materials to maximise process efficiency and reduce costs. Research projects have ranged from fundamental properties of feed minerals to feasibility studies of possible implantation on a national scale.

Ron Zevenhoven has been working on CO_2 mineral sequestration since 2000, initially at (current) Aalto University in Espoo and since 2005 at Åbo Akademi University in Turku, Finland. He has (co-)supervised eight PhD theses on the subject and (co-) authored quite a few peer-reviewed papers and book chapters. He has an MSc and PhD (Chem Eng) from Delft University of Technology and, in Finland since 1993, holds the professorship in Engineering Thermodynamics and Modelling at ÅAU since 2005. He was one of the authors of the 2005 IPCC Special Report on Carbon Dioxide Capture and Storage and was Chairman of the organising and scientific committees of ACEME2010 (http://web.abo.fi/fak/tkf/vt/aceme10).

Dr Inês S. Romão completed her PhD thesis titled "Production of magnesium carbonates from serpentinites for CO_2 mineral sequestration — optimisation towards industrial application" in 2015 as a double-degree at Åbo Akademi University in Turku, Finland and the University of Coimbra, Portugal under the supervision of Profs Ron Zevenhoven and Licínio Gando Ferreira. She currently lives in Switzerland.

Contents

**Chapter 3.   CO$_2$ Mineralisation as a Route
to Energy-Efficient CO$_2$ Sequestration
or Materials with Market Value   41**

Ron Zevenhoven and Inês S. Romão

Chapter 1

Mineral Carbonation Technology Overview

Qi Liu[*,†], M. Mercedes Maroto-Valer[†] and Aimaro Sanna[†]
*Research Institute of Enhanced Oil Recovery,
China University of Petroleum, Beijing 102249, China
†Centre for Innovation in Carbon Capture and Storage,
School of Engineering and Physical Sciences,
Heriot-Watt University, Edinburgh, EH14 4AS, UK

1.1 Introduction

Increasing atmospheric greenhouse gas (GHG) emissions are widely accepted as the main contributor of climate change. Carbon dioxide (CO_2) is the main anthropogenic GHG and represents 76% of total global GHGs (Edenhofer et al., 2014; IPCC, 2014). Climate change has been proven to be in large part caused by human activities and confirmed by observation of increased global average surface temperatures, rising global average sea levels and widespread melting of Northern Hemisphere snow cover (IPCC, 2007; Smith et al., 2009). CO_2 levels in the atmosphere are therefore increasing at an alarming rate, and failing to cut atmospheric CO_2 emissions significantly during the 21st century may result in a series of calamitous environmental consequences (Stern, 2006). The latest United Nations Climate Change Conference (COP21) held in Paris in 2015 delivered a breakthrough on the response to climate change from the international community. The result of the conference was a new international agreement on climate change to keep global warming below 2°C (COP21, 2015).

There is a wide portfolio of potential technologies to mitigate CO_2 emissions (Herzog, 2001; Mazzotti *et al.*, 2009; IPCC, 2014), including: (i) enhancing energy efficiency, for example improving turbines; (ii) switching to less carbon-intensive fossil fuels, such as using natural gas as main fuel instead of coal; (iii) increasing use of low- and near-zero-carbon energy sources such as solar, wind, hydro, nuclear, biomass, geothermal and tidal power and (iv) CO_2 capture, storage and utilisation (CCUS) including options such as geological, ocean, terrestrial, advanced chemical approaches and advanced biological processes (IPCC, 2014). CCUS is a promising option for stabilising atmospheric CO_2 concentration (IPCC, 2005, 2014). CCUS involves capturing CO_2 from large point sources, transporting and storing it permanently, so it will not enter the atmosphere. One potentially attractive approach for CCUS is injection of CO_2 into underground formations, such as active and depleted oil and gas reservoirs, deep unmineable coal seams and deep saline aquifers (Bachu, 2008; Liu and Maroto-Valer, 2011). However, this option is still lacking demonstration on large power plants and it is not a valid option in numerous locations worldwide.

Mineral carbonation (MC) is emerging as a potential CCUS technology solution to sequester CO_2 from small and medium emitters in many regions, especially where CO_2 underground sequestration is not a viable option (Zevenhoven *et al.*, 2011). MC involves the reaction of CO_2 with metal oxide (like magnesium oxide (MgO) and calcium oxide (CaO)) bearing materials to form stable calcium/magnesium carbonates. This naturally occurring process called silicate weathering takes place on a geological time scale (IPCC, 2005; Sanna *et al.*, 2014).

Ex-situ MC is principally translating this very slow natural weathering process into economically viable technology, which was first suggested as a CCSU method by Seifritz in 1990 (Seifritz, 1990). There are two main advantages making MC unique amongst the other storage options, (i) CO_2 is stored in a stable solid form (carbonate rock) and (ii) the abundance of metal oxide bearing materials, particularly of natural silicates (Olajire, 2013). However, the deployment of MC technologies is still hindered by a number of reasons such as the slow reaction kinetics, the large material flows involved and concerns about process energy requirements (Zevenhoven and Fagerlund, 2010). Recent developments include an increasing number of research papers and patents and a trend towards scale-up and commercial demonstration in the coming decades (Olajire, 2013; Sanna *et al.*, 2014).

1.2 Process Principles

MC can be defined as the reaction of metal oxide bearing materials (indicated here as MO, where M is a divalent metal, e.g. calcium, magnesium or iron) with CO_2 to form the corresponding insoluble carbonates and releasing heat, with the general chemical reaction shown in Reaction (1.1) (IPCC, 2005):

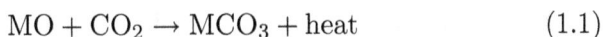

$$MO + CO_2 \rightarrow MCO_3 + \text{heat} \qquad (1.1)$$

This reaction can take place either below (*in-situ*) or above (*ex-situ*) ground. *In-situ* MC is the process where CO_2 is directly injected into underground reservoirs to promote the reaction between CO_2 and alkaline minerals in the geological formation (IPCC, 2005; Sanna *et al.*, 2014); while *ex-situ* MC relates to above-ground processes, which require rock mining and material comminution as MC process pre-requisites (IPCC, 2005).

The amount of heat released in Reaction (1.1) depends on the specific metal and on the material containing the metal oxide (see Table 3.1). Generally, this is a large fraction (up to 46% in the case of CaO) of the heat released by the upstream combustion process forming CO_2 ($393.8\,\text{kJmol}^{-1}CO_2$ for combustion of elemental carbon). Carbonation of natural silicate minerals that contain alkaline-earth oxides like MgO and CaO traps CO_2 as environmentally stable solid carbonates, which for serpentine (Reaction (1.2)), olivine (Reaction (1.3)) and wollastonite (Reaction (1.4)) minerals can be described by the given exothermic reactions (Robie *et al.*, 1978; Lackner, 2002). Heat values are given per unit mol of CO_2 at standard conditions of

25°C and 0.1 MPa in all cases.

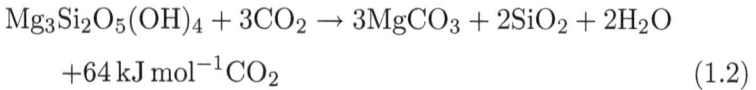

$$Mg_3Si_2O_5(OH)_4 + 3CO_2 \rightarrow 3MgCO_3 + 2SiO_2 + 2H_2O$$
$$+64\,kJ\,mol^{-1}CO_2 \tag{1.2}$$

$$MgSiO_3 + CO_2 \rightarrow MgCO_3 + SiO_2 + 89\,kJ\,mol^{-1}CO_2 \tag{1.3}$$

$$CaSiO_3 + CO_2 \rightarrow CaCO_3 + SiO_2 + 90\,kJ\,mol^{-1}CO_2 \tag{1.4}$$

These minerals contain MgO and CaO at concentrations between 30 and 50 wt.%. However, the calcium-based resources are much less common than the magnesium-based (see Chapter 2), so large-scale MC primarily based on calcium bearing silicates is unlikely (Zevenhoven *et al.*, 2011). Apart from these three metal oxide bearing minerals mentioned above, the abundance of materials like basalt with 10–20 wt.% of MgO and CaO content has been considered as promising material for MC process (McGrail *et al.*, 2006; Schaef *et al.*, 2009). The above topics are treated in Chapter 2.

The formation of carbonates is thermodynamically favoured at low temperature, as the reactions release heat. While their reverse reactions (calcination) are favoured at high temperature (above 800°C for calcium carbonate and above 300°C for magnesium carbonate, at a CO_2 partial pressure of one bar). Despite the fact that carbonation of metal oxide bearing minerals occurs spontaneously on geological time scales even at the low partial pressure of atmospheric CO_2 and at ambient temperature (Robie *et al.*, 1978; Lasaga and Berner, 1998), the urgent need for reducing the atmospheric CO_2 to the pre-industrial level before 2100, implies that we need to engineer the natural MC to achieve fast reaction rates. Therefore, one of the requirements for MC to become economically viable is to significantly accelerate the process (Krevor and Lackner, 2009). Moreover, the heat of reaction within the environmental constraints such as with minimal energy and material losses, need to be exploited, to minimise the energy consumed in the process (IPCC, 2005). The fact that the overall chemistry of Reaction (1.1) is exothermic implies that proper optimisation and process integration should allow for operation at zero or negative net energy input. However, if these stages operate at different temperatures, pressures and energy requirements, the overall process may need energy input (Zevenhoven and Fagerlund, 2010). Nevertheless, mineral resource availability, scalability, applicability to regions without geologic storage capacity, inherent stability of the reaction products and the potential revenue from MC products promote the ongoing development of this technology (Sanna *et al.*, 2014).

1.3 Resources for MC

MC has a large CO_2 sequestration potential ($>10,000$ Gt C) due to the large abundance of resources around the world, as shown in Figure 1.1 (see also Figures 2.10, 2.11 and 3.1) (Sanna *et al.*, 2014). Resources for MC focus on metal oxide bearing minerals containing alkaline-earth metals (such as calcium and magnesium) as opposed to

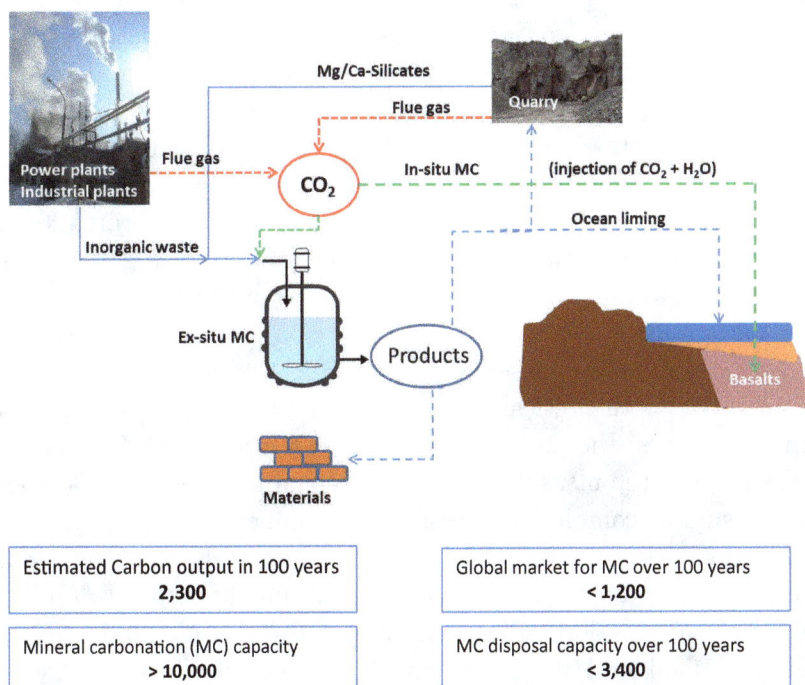

Estimated Carbon output in 100 years **2,300**	Global market for MC over 100 years **< 1,200**
Mineral carbonation (MC) capacity **> 10,000**	MC disposal capacity over 100 years **< 3,400**

Figure 1.1 Overview of MC process, MC capacity, estimated carbon output in 100 years and MC disposal capacity. Data in Gt C (adapted from Sanna *et al.*, 2014).

alkali metals (such as sodium and potassium) because alkali metals corresponding carbonates are very soluble in water (IPCC, 2005). Potential feedstock minerals are available as a result of large sections of ancient ocean floor that have been thrust on to certain continental margins as a result of continental–oceanic crustal plate collisions, which left portions of ocean floor at the surface of continental crust (Voormeij and Simandl, 2004). The western coast of North America and the eastern United States contain large amounts of this material (Olajire, 2013). Oxides and hydroxides of Ca and Mg would be good sources for MC, but they are extremely rare in nature due to their reactivity. However, these two elements are most commonly found in silicate minerals (e.g. olivine, serpentine), which can also be carbonated, as carbonic acid (H_2CO_3; pKa $= 6.3$) is stronger than silicic acid ($Si(OH)_4$; pKa $= 9.50$) (Olajire, 2013). Therefore, suitable metal oxide bearing minerals may be silicate rocks because of their occurrence as large deposits at numerous locations worldwide.

Mafic and ultramafic rocks are silicates containing high amounts of magnesium, calcium and iron and have a low content of sodium and potassium. Olivine, serpentine, enstatite ($MgSiO_3$), talc ($Mg_3Si_4O_{10}(OH)_2$) and wollastonite are the main mineral constituents in mafic and ultramafic rocks (IPCC, 2005; Olajire, 2013). The carbonation of magnesium-based silicate minerals gives by far the highest capacity and longest storage time of the currently known CCS options (Lackner, 2003). The distribution and evaluation of feedstocks for MC are described in detail in Chapter 2.

In order to theoretically compare the CO_2 binding capacity of mineral sources, the concept of RCO_2 was introduced. The RCO_2 is the ratio of the mass of mineral needed to the mass of CO_2 fixed when assuming complete conversion of the mineral upon carbonation (Goff and Lackner, 1998). The typical values of RCO_2 range from 1.8 to 3 tonnes of mineral/tCO_2, which mainly depends on the degree of carbonation conversion (Zevenhoven *et al.*, 2011). RCO_2 values (as t mineral/tC) range from 1.6 tonnes of mineral/tCO_2 for forsterite, 2.1 for serpentine, 2.4 for basalt, 2.6 for wollastonite to 6.3 for anorthite (calcium aluminium silicate) (Oelkers *et al.*, 2008). Serpentine and olivine are mainly found in ophiolite belts, where colliding continental plates caused an uplifting of the earth's crust (Coleman, 1977). For instance, RCO_2 values have been reported between 1.97 and 2.51 (depending on purity and type) in the Eastern United States and in Puerto Rico, where ultramafic deposits contain serpentine and olivine (IPCC, 2005).

It is important to distinguish between *in-situ* and *ex-situ* MC methods, where the former explicitly aim at reacting the CO_2 to form carbonates with alkaline minerals present in a geological formation without moving any rock material (Zevenhoven and Fagerlund, 2010). Important applications for *in-situ* carbonation are found in Iceland and Oman (Matter and Kelemen, 2008; Oelkers *et al.*, 2008). For above-ground (*ex-situ*) processes, a metric tonne of CO_2 will require 2.5–3 tonnes of magnesium silicate mineral (Olajire, 2013). Although, *in-situ* MC is generally discussed in Chapter 2, this book is dedicated to the *ex-situ* MC processes, which are discussed in depth in all the remnant chapters.

In terms of small-scale deployment, carbonation of waste materials, ashes and industrial by-products is rapidly expanding as they can provide sources of alkalinity, with the benefit of binding significant amounts of CO_2 and/or producing a valuable carbonate material. Waste streams of calcium silicate materials that have been considered for MC include pulverised fuel ash from coal-fired power plants (with a CaO content up to 65 wt.%), bottom ash (about 20 wt.% CaO) and fly ash (about 35 wt.% CaO) from municipal solid waste incinerators, de-inking ash from paper recycling (about 35% by weight CaO), stainless steel slag (about 65 wt.% of CaO and MgO) and waste cement (Johnson, 2000; Bertos *et al.*, 2004, Iizuka *et al.*, 2004). The global potential of calcium-containing waste materials is limited to binding a few 100 Mt/a^3. Even though their total binding amounts are too small to substantially reduce CO_2 emissions, it can be of great importance to small emitters like iron or steel plants. An overall benefit would then be the development of technical infrastructure for MC, as they could help introduce this technology to increase public acceptance (Bobicki *et al.*, 2012; Geerlings and Zevenhoven, 2013). MC processes based on inorganic wastes are extensively discussed in Chapter 4.

1.4 *Ex-Situ* MC Processes

Ex-situ MC can be conducted either via direct or indirect processes (Sanna *et al.*, 2014). Direct MC is the simplest approach, where the carbonation of the mineral takes place in a single process step through direct gas–solid MC or direct aqueous MC (Bobicki *et al.*, 2012). Direct gas–solid MC is the most basic form of direct MC, where CO_2 reacts with minerals to produce carbonates (Gupta and Fan, 2002; Lee *et al.*, 2008; Baciocchi *et al.*, 2009; Lim *et al.*, 2010; Azdarpour *et al.*, 2015). Direct aqueous MC is the reaction of CO_2 with minerals in an aqueous solution in a single step (Chen *et al.*, 2006; Alexander *et al.*, 2007; Krevor and Lackner, 2009; Liang *et al.*, 2009; Prigiobbe *et al.*, 2009; Azdarpour *et al.*, 2015). A detailed evaluation and classification of the *ex-situ* MC technologies is provided in Chapters 3 and 4.

Overall, direct routes present straightforward design and the absence of non-aqueous solvents (Sanna *et al.*, 2014). However, the reaction conversion rates are low and high CO_2 pressure and temperature are also required (Sipilä *et al.*, 2005). In order to increase reaction conversion rates and efficiencies, a variety of pre-treatments have been developed, including high energy mechanical grinding and chemical leaching, while other methods such as thermal and mechano-chemical pre-treatments have also been developed (Gerdemann *et al.*, 2007; Balucan *et al.*, 2013; Declercq *et al.*, 2013; Sanna *et al.*, 2014). Figure 1.2 (also see Table 3.2) shows the direct and indirect *ex-situ* MCs.

Indirect MC consists of two or more stages (Sanna *et al.*, 2014), including the extraction of reactive components (Mg^{2+}, Ca^{2+}) from feedstock by using acids or other solvents, followed by the precipitation of carbonate materials, which takes place as separate step and/or in different reactors. The end products of indirect MC are usually of higher purity compared to those obtained by direct-processes, due to the removal of impurities in intermediate steps (Zevenhoven and Fagerlund, 2010; Bobicki *et al.*, 2012; Eloneva *et al.*, 2012; Olajire, 2013).

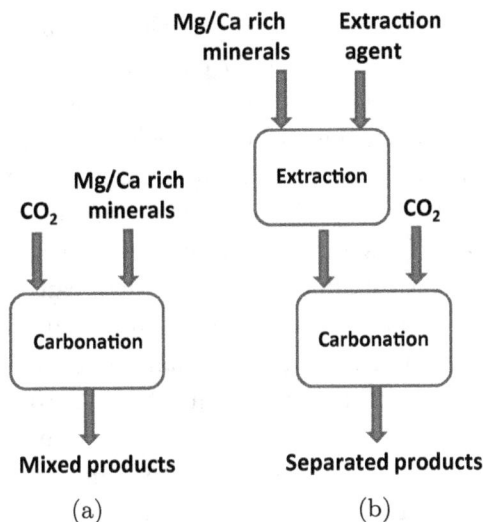

Figure 1.2 A schematic representation of the principles of direct and *ex-situ* indirect MCs. (a) Direct process (one step). (b) Indirect process (two or more steps) (adapted from Bobicki *et al.*, 2012).

Indirect MC can be accomplished through different process routes, such as indirect multi-stage gas–solid MC, pH swing processes, HCl extraction, molten salt process, other acid extractions, bioleaching, ammonia extraction and caustic extraction (Bobicki *et al.*, 2012; Azdarpour *et al.*, 2015). The advantages and challenges of these technologies are discussed in Chapters 3–5.

The aqueous MC rate is higher compared to the gas–solid reaction using NaOH, NaCl, $NaHCO_3$, $KHCO_3$, etc. (Sanna *et al.*, 2014). Indirect aqueous carbonation consists of two aqueous separate steps for the extraction and the carbonation of Ca and/or Mg. The advantage of this route is that the two steps can be optimised by using additives separately, incorporating additional steps, if needed (Olajire, 2013). Multi-step aqueous indirect processes in the presence of additives are able to reach high carbonation efficiency using mild process conditions and short residence time as a result of faster reaction kinetics in the presence of additives. However, the energy intensive chemical regeneration step is slowing the development of this group of technologies (Sanna *et al.*, 2014).

From a technical point of view, direct MC is less viable on large industrial scale at the current scale of development due to its current constrains, such as the dissolution of minerals, product layer diffusion control to reduce the CO_2 diffusion or carbonate precipitation and CO_2 dissolution. MC is more favoured via the indirect route for industrial emitters, as higher purity products can be produced. In addition, the conversion rate of calcium and magnesium to carbonates is significantly higher for indirect than direct processes (Olajire, 2013; Assima *et al.*, 2014; Azdarpour *et al.*, 2015). In particular, the potential production of sellable products (see Figure 1.1), the co-removal of pollutants from the flue gas and process integration are essential aspects to lower the costs (Sanna *et al.*, 2014). *Ex-situ* MC could be retrofit to existing small or medium emitters due to the requirement to handle large amount of mineral feedstock and reaction products; while *ex-situ* MC may be suitable only to new designed large emitters, which would require dedicated additional infrastructures. *Ex-situ* MC may target small–medium emitters, as geological sequestration is unlikely to be an economically viable option. Researchers have reported that South Africa, China, Russia, Kazakhstan, Australia, USA and Europe have large ultramafic rock deposits within a 100–200 km radius of power/industrial plants which emits over 1 Mt per year CO_2 (Wee, 2013; Sanna *et al.*, 2014). However, not all these resources are easily accessible.

Apart from natural resources, various large-scale industrial wastes can be considered as feedstock for CO_2 mineralisation (Chapter 4). The current MC technologies developed for wastes still cannot compete with geological storage in terms of potential quantity and cost of sequestrated CO_2, even though it has several benefits, such as avoiding costs for mining and transport. These options should be considered in countries that lack CO_2 geological storage formations (Mai *et al.*, 2009).

1.4.1 *MC Process Scale and Product Applications*

If MC process technologies are to be widely implemented on a commercial scale, they have to satisfy several criteria, including versatility and practicality, scalability, economics and energy consumption of the process, as well as quality, quantity and marketability of the products have to be taken into consideration.

Chapter 5 focuses on discussing the above aspects of MC, and how these aspects can be expanded upon to strengthen the case for the development and widespread implementation of MC.

The effective development of utilisation routes for the MC products could help to make this technology economically viable and facilitate its deployment. Applications of MC products can be divided into low-end high-volume and high-end low-volume uses. For the MC products to be commercially used, there are specifications and quality criteria that must be met. Construction and filler applications seem to be the most appropriate for silica and carbonate products, respectively, while feedstock for iron/steel works may represent the natural pathway for iron oxides from MC. Among the low-end applications, land reclamation from the sea in coastal areas and mine reclamation using silica, magnesium and calcium carbonates are other possible low-tech high-volume applications (Sanna *et al.*, 2014).

Finally, Chapter 5 discusses the potential integration of direct and indirect MC techniques in a hybrid process, showing how the two segments can complement each other by bringing mutual benefits to the overall integrated process.

1.5 Conclusions

Ex-situ MC of CO_2 is an important CCS approach that provides an alternative for CO_2 storage in underground formations. It can provide leakage-free CO_2 fixation that does not require post-storage monitoring, and has an overwhelmingly large worldwide capacity in terms of available feedstocks. Moreover, the solid products can be used in applications ranging from land reclamation to iron and steelmaking. *Ex-situ* MC has been demonstrated, but its widespread application is currently limited by its high costs. Energy use, the reaction rate and material handling are the main issues hindering the deployment of this technology. The value of the products seems central to render *ex-situ* MC economically viable; similarly conventional geological storage would be profitable only when combined with enhanced oil recovery. In terms of *in-situ* MC, it can be a very promising option due to resources available and enhanced security, but the technology is still in its infancy and transport and storage costs are still higher than geological storage in sedimentary basins.

References

Alexander, G., M. Maroto-Valer and P. Gafarova-Aksoy (2007). Evaluation of reaction variables in the dissolution of serpentine for mineral carbonation. *Fuel* 86(1–2): 273–281.

Assima, G. P., F. Larachi, J. Molson and G. Beaudoin (2014). Comparative study of five Québec ultramafic mining residues for use in direct ambient carbon dioxide mineral sequestration. *Chemical Engineering Journal* 245: 56–64.

Azdarpour, A., M. Asadullah, E. Mohammadian, H. Hamidi, R. Junin and M. A. Karaei (2015). A review on carbon dioxide mineral carbonation through pH-swing process. *Chemical Engineering Journal* 279: 615–630.

Bachu, S. (2008). CO₂ storage in geological media: Role, means, status and barriers to deployment. *Progress in Energy and Combustion Science* 34(2): 254–273.

Baciocchi, R., G. Costa, A. Polettini, R. Pomi and V. Prigiobbe (2009). Comparison of different reaction routes for carbonation of APC residues. *Energy Procedia* 1(1): 4851–4858.

Balucan, R. D., B. Z. Dlugogorski, E. M. Kennedy, I. V. Belova and G. E. Murch (2013). Energy cost of heat activating serpentinites for CO₂ storage by mineralisation. *International Journal of Greenhouse Gas Control* 17(5): 225–239.

Bertos, M. F., S. J. R. Simons, C. D. Hills and P. J. Carey (2004). A review of accelerated carbonation technology in the treatment of cement-based materials and sequestration of CO(2). *Journal of Hazardous Materials* 112(3): 193–205.

Bobicki, E. R., Q. Liu, Z. Xu and H. Zeng (2012). Carbon Capture and Storage Using Alkaline Industrial Wastes. *Fuel & Energy Abstracts* 38(2): 302–320.

Chen, Z.-Y., W. K. O'Connor and S. J. Gerdemann (2006). Chemistry of aqueous mineral carbonation for carbon sequestration and explanation of experimental results. *Environmental Progress* 25(2): 161–166.

Coleman, R. G. (1977). *Ophiolites*, Springer-Verlag, Berlin.

COP21 (2015). ParisAgreement: Points that remain in suspense. Retrieved 09/02/2016. Available from: http://www.cop21.gouv.fr/en/.

Declercq, J., O. Bosc and E. H. Oelkers (2013). Do organic ligands affect forsterite dissolution rates? *Applied Geochemistry* 39(8): 69–77.

Edenhofer, O., R. Pichs-Madruga, Y. Sokona, S. Kadner, J. C. Minx, S. Brunner, S. Agrawala, G. Baiocchi, I. A. Bashmakov, G. Blanco, J. Broome,

T. Bruckner, M. Bustamante, L. Clarke, M. Conte Grand, F. Creutzig, X. Cruz-Núñez, S. Dhakal, N. K. Dubash, P. Eickemeier, E. Farahani, M. Fischedick, M. Fleurbaey, R. Gerlagh, L. Gómez-Echeverri, S. Gupta, J. Harnisch, K. Jiang, F. Jotzo, S. Kartha, S. Klasen, C. Kolstad, V. Krey, H. Kunreuther, O. Lucon, O. Masera, Y. Mulugetta, R. B. Norgaard, A. Patt, N. H. Ravindranath, K. Riahi, J. Roy, A. Sagar, R. Schaeffer, S. Schlömer, K. C. Seto, K. Seyboth, R. Sims, P. Smith, E. Somanathan, R. Stavins, C. von Stechow, T. Sterner, T. Sugiyama, S. Suh, D. Ürge-Vorsatz, K. Urama, A. Venables, D. G. Victor, E. Weber, D. Zhou, J. Zou and T. Zwickel (2014). Technical Summary. In: *Climate Change 2014: Mitigation of Climate Change. Contribution of Working Group III to the Fifth Assessment Report of the Intergovernmental Panel on Climate Change* [Edenhofer, O., R. Pichs-Madruga, Y. Sokona, E. Farahani, S. Kadner, K. Seyboth, A. Adler, I. Baum, S. Brunner, P. Eickemeier, B. Kriemann, J. Savolainen, S. Schlömer, C. von Stechow, T. Zwickel and J. C. Minx (eds.)]. Cambridge University Press, Cambridge, United Kingdom and New York, NY, USA.

Eloneva, S., A. Said, C. J. Fogelholm and R. Zevenhoven (2012). Preliminary assessment of a method utilizing carbon dioxide and steelmaking slags to produce precipitated calcium carbonate. *Applied Energy* 90(1): 329–334.

Geerlings, H. and R. Zevenhoven (2013). CO_2 mineralization-bridge between storage and utilization of CO_2. *Annual Review of Chemical & Biomolecular Engineering* 4(3): 103–117.

Gerdemann, S. J., W. K. O'Connor, D. C. Dahlin, L. R. Penner and R. Hank (2007). *Ex situ* aqueous mineral carbonation. *Environmental Science & Technology* 41(41): 2587–2593.

Goff, F. and K. S. Lackner (1998). Carbon dioxide sequestering using ultramafic rocks. *Environmental Geosciences* 5(3): 89–102.

Gupta, H. and L. S. Fan (2002). Carbonation-calcination cycle using high reactivity calcium oxide for carbon dioxide separation from flue gas. *Industrial & Engineering Chemistry Research* 41(16): 4035–4042.

Herzog, H. J. (2001). Peer reviewed: What future for carbon capture and sequestration? *Environmental Science & Technology* 35(7): 148A–153A.

Iizuka, A., M. Fujii, A. Yamasaki and Y. Yanagisawa (2004). A new CO_2 sequestration process via carbonation of waste cement. *Industrial & Engineering Chemistry Research* 43: 7880–7887.

IPCC (2005). *IPCC Special Report on Carbon Dioxide Capture and Storage.* Cambridge, UK.

IPCC (2007). *Climate Change 2007: The Physical Science Basis.* Fourth Assessment Report. IPCC (Intergovernmental Panel on Climate Change), Geneva.

IPCC (2014). *Climate Change 2014: Synthesis Report.* Contribution of Working Groups I, II and III to the Fifth Assessment Report of the Intergovernmental Panel on Climate Change [Core Writing Team, R.K. Pachauri and L.A. Meyer (eds.)]. IPCC, Geneva, Switzerland, p. 151.

Johnson, D. C. (2000). Accelerated carbonation of waste calcium silicate materials. *SCI Lecture Papers Series* 108/2000: 1–10.

Krevor, S. C. and K. S. Lackner (2009). Enhancing process kinetics for mineral carbon sequestration. *Energy Procedia* 1(1): 4867–4871.

Lackner, K. S. (2002). Carbonate chemistry for sequestering fossil carbon. *Annual Review of Energy and the Environment* 27(1): 193–232.

Lackner, K. S. (2003). Climate change. A guide to CO_2 sequestration. *Science* 300(5626): 1677–1678.

Lackner, K. S., C. H. Wendt, D. P. Butt, E. L. Joyce Jr and D. H. Sharp (1995). Carbon dioxide disposal in carbonate minerals. *Energy* 20(11): 1153–1170.

Lasaga, A. C. and R. A. Berner (1998). Fundamental aspects of quantitative models for geochemical cycles. *Chemical Geology* 145(3–4): 161–175.

Lee, S. C., H. J. Chae, S. J. Lee, B. Y. Choi, C. K. Yi, J. B. Lee, C. K. Ryu and J. C. Kim (2008). Development of regenerable MgO-based sorbent promoted with K_2CO_3 for CO_2 capture at low temperatures. *Environmental Science & Technology* 42(8): 2736–2741.

Liang, Z., S. Liqin, J. Junfeng and T. H. Henry (2009). Aqueous carbonation of natural brucite: Relevance to CO_2 sequestration. *Environmental Science & Technology* 44(1): 406–411.

Lim, M., G.-C. Han, J.-W. Ahn and K.-S. You (2010). Environmental remediation and conversion of carbon dioxide (CO_2) into useful green products by accelerated carbonation technology. *International Journal of Environmental Research and Public Health* 7(1): 203.

Liu, Q. and M. Maroto-Valer (2011). Parameters affecting mineral trapping of CO_2 sequestration in brines. *Greenhouse Gases Science & Technology* 1(3): 211–222.

Mai, U., U. Mati and K. Rein (2009). CO_2 mineral sequestration in oil-shale wastes from Estonian power production. *Journal of Environmental Management* 90(2): 1253–1260.

Matter, P. B. and J. Kelemen (2008). From the cover: *In situ* carbonation of peridotite for CO_2 storage. *Proceedings of the National Academy of Sciences of the United States of America* 105(45): 17295–17300.

Mazzotti, M., R. Pini and G. Storti (2009). Enhanced coalbed methane recovery. *The Journal of Supercritical Fluids* 47(3): 619–627.

McGrail, B. P., H. T. Schaef, A. M. Ho, Y. J. Chien, J. J. Dooley and C. L., Davidson (2006). Potential for carbon dioxide sequestration in flood basalts. *Journal of Geophysical Research* 111(B12201): 1–13.

Oelkers, E. H., S. R. Gislason and J. Matter (2008). Mineral carbonation of CO_2. *Elements* 4(5): 333–337.

Olajire, A. A. (2013). A review of mineral carbonation technology in sequestration of CO_2. *Journal of Petroleum Science & Engineering* 109(5): 364–392.

Prigiobbe, V., M. Hänchen, M. Werner, R. Baciocchi and M. Mazzotti (2009). Mineral carbonation process for CO_2 sequestration. *Energy Procedia* 1(1): 4885–4890.

Robie, R. A., B. S. Hemmingway and J. R. Fisher (1978). Thermodynamic properties of minerals and related substances at 298. 15 K and 1 bar (10/sup 5/ pascals) pressure and at higher temperatures. US Geological Bulletin 1452, Washington DC.

Sanna, A., M. Uibu, G. Caramanna, R. Kuusik and M. Maroto-Valer (2014). A review of mineral carbonation technologies to sequester CO_2. *Chemical Society Reviews* 43(23): 8049–8080.

Schaef, H. T., B. P. McGrail and A. T. Owen (2009). Basalt- CO_2–H_2O interactions and variability in carbonate mineralization rates. *Energy Procedia* 1(1): 4899–4906.

Seifritz, W. (1990). CO_2 disposal by means of silicates. *Nature* 345(6275): 486–486.

Sipilä, J., S. Teir and R. Zevenhoven (2005). Carbon dioxide sequestration by mineral carbonation — Literature review update 2005–2007. *Report Vt* 3(8): 13.

Smith, J. B., S. H. Schneider, M. Oppenheimer, G. W. Yohe, W. Hare, M. D. Mastrandrea, A. Patwardhan, I. Burton, J. Corfee-Morlot, C. H. D. Magadza, H.-M. Füssel, A. B. Pittock, A. Rahman, A. Suarez and J.-P. van Ypersele (2009). Assessing dangerous climate change through an update of the Intergovernmental Panel on Climate Change (IPCC) "reasons for concern". *Proceedings of the National Academy of Sciences* 106(11): 4133–4137.

Stern, N. (2006). *The Economics of Climate Change: The Stern Review.* Cambridge University Press, Cambridge, UK.

Teir, S., S. Eloneva, C.-J. Fogelholm and R. Zevenhoven (2009). Fixation of carbon dioxide by producing hydromagnesite from serpentinite. *Applied Energy* 86(2): 214–218.

Voormeij, D. A. and G. J. Simandl (2004). Ultramafic rocks in British Columbia: delineating targets for mineral sequestration of CO_2. British Columbia Ministry of Energy and Mines, Resource Development and Geoscience Branch.

Wee, J.-H. (2013). A review on carbon dioxide capture and storage technology using coal fly ash. *Applied Energy* 106: 143–151.

Zevenhoven, R. and J. Fagerlund (2010). Mineralisation of carbon dioxide. In: *Developments and Innovation in CCS Technology* [Maroto-Valer, M. (ed.)]. Woodhead Publishing Ltd, Cambridge, vol. 2, pp. 433–462.

Zevenhoven, R., J. Fagerlund and J. K. Songok (2011). CO_2 mineral sequestration: Developments toward large-scale application. *Greenhouse Gases: Science and Technology* 1(1): 48–57.

Chapter 2

Rocks: Cation Donors
with Enormous Resources

Mike T. Styles and Alicja M. Lacinska
British Geological Survey, Keyworth, Nottingham, UK

2.1 Introduction

The carbon capture and storage by mineralisation (CCSM) process requires a large stock of feed material that will, after appropriate processing, provide a source of cations for reaction with carbon dioxide (CO_2) to form stable carbonate. Limited amounts of the material can come from processing waste products however, for implementation on a large scale, most must come from rocks. Over millions of years, the weathering and breakdown of rocks created products that have played a vital role in the generation and balance of CO_2 in the earth's atmosphere. This process can be encouraged, enhanced and engineered to play an important role in reducing the rising levels of atmospheric CO_2.

Two types of CCSM have been proposed: *ex-situ* and *in-situ*, both having slightly different requirements. The *ex-situ* mineralisation will need large amounts of feed material to be moved to a processing plant and, after carbonation, the end products moved away. The processing is required to be fast, probably a few hours or ideally much less, and effective, i.e. using materials with a high content of easily extractable cations ready for carbonation. *In-situ* mineralisation will involve injection into large rock bodies and hence both a lower total reactive cation content and slower reaction rates (years) could be tolerable, as long as carbonation takes place before the CO_2 migrates back to the surface. Hence, many possible resources additional to those suitable for *ex-situ* mineralisation may be available.

2.2 Composition and Classification of Ultramafic and Mafic Rocks

The mineral carbonation technology will require large amounts of cations that are suitable for forming stable (insoluble in water) carbonates. The earth's crust is largely composed of silicate minerals with abundances being dominated by silica (SiO_2 ∼60.0%) and alumina (Al_2O_3 ∼16%) with much lower amounts of all the other elements. The commonest crustal element potentially used for carbonation is calcium (CaO 6.4%), but by far the most abundant Ca-rich rock is limestone, where the Ca is already present as carbonate. Other Ca-rich rocks potentially suitable for CCSM are rare. The alkali metals form soluble carbonates and hence are not suitable. Sources for *ex-situ* CCSM from rocks will largely have to be magnesium (MgO 4.7%) and/or iron (FeO 6.7%), however the latter is a "valuable" metal for industrial uses and FeO-rich rocks are used as ores.

All the CCSM processes are based around a simplified chemical exchange, and in the rock-based system, the CO_2 capture can be exemplified by the transition from a divalent metal silicate to a divalent metal carbonate:

$$\text{Mg silicate} + CO_2 \rightarrow \text{Mg carbonate} + SiO_2 \tag{2.1}$$

Most rocks contain ≥ 50% silica that makes them poor targets for carbonation. Higher proportions of the desired cations are found in crystalline, Mg, Fe-rich minerals and igneous rocks. These rocks are classified as mafic when they contain ∼25% MgO + CaO + FeO and ∼50% SiO_2 or ultramafic with ∼45% MgO + FeO + CaO and < 45% SiO_2. The latter are clearly the best target for mineralisation and are likely to be the key resource for industrial scale CCSM; their composition and classification is described in the following section. The mafic rocks may also be a possible host for *in-situ* mineralisation.

2.3 Ultramafic Rocks

Ultramafic rocks are composed of $> 90\%$ mafic minerals in various proportions with trace amounts of other phases. The main mafic minerals are the Mg-silicates: olivine (Mg_2SiO_4), orthopyroxene ($MgSiO_3$) and clinopyroxene (($MgCa$)SiO_3). In most rocks, a proportion of the Mg in the crystal structure of these minerals is substituted by Fe^{2+}. In the ultramafic rocks, maximum substitution is only around 10% and for the sake of simplicity, only the names of the mineral or the Mg end-member will be used.

Ultramafic rocks are formed by the crystallisation of an anhydrous, Mg-rich silicate melt at temperatures around 1,200°C. This usually occurs in the middle or lower parts of the earth's crust or in the earth's mantle. Figure 2.1 shows the internationally accepted classification for fresh ultramafic rocks with a slight modification after the British Geological Survey, with the names used for the rocks and the proportions of the various minerals. The apices of the

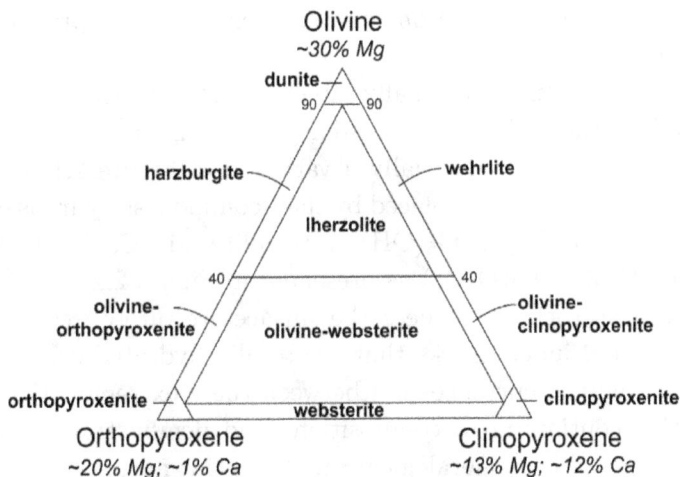

Figure 2.1 The classification diagram of fresh, anhydrous ultramafic rocks. The Mg and Ca content of the rocks at each apex expressed in wt.% element (from Styles *et al.*, 2014).

triangle show rocks composed of 100% of a particular mineral, and the inner parts of the triangle show various proportionate mixtures of the minerals. The Mg and Ca contents of the different minerals are indicated. The diagram clearly shows that rocks rich in olivine are the best target as they have the highest proportion of divalent cation and thus the lowest silica content. Rocks consisting dominantly of olivine are known as peridotites, whereas those in which pyroxene is dominant are termed pyroxenites.

Following their formation at depth, the ultramafic rocks may be brought up towards the earth's surface during tectonic processes and erosion lasting millions of years. In several localities, such complexes are now exposed at the surface or at shallow depths making them potentially available as a source material for CCSM. Rocks that were in equilibrium in an anhydrous, high temperature environment become highly unstable during this transport to the surface and final residence, and consequently undergo a range of mineralogical changes. These changes are principally due to the ingress of hydrous fluids present at shallower depths, causing mineral hydration. The interaction with hydrous fluids rarely happens as a single-stage alteration and the resultant rocks may contain mineral assemblages reflecting partial equilibration at several temperatures, pressures and fluid compositions.

There is no internationally accepted classification of altered ultramafic rocks or indeed consistency in the names used. Here we use a scheme that is essentially a variation of Figure 2.1, but with the anhydrous minerals replaced by their commonest hydrous equivalents: serpentine ($Mg_3Si_2O_5(OH)_4$), tremolite ($Mg_5Ca_2Si_8O_{22}(OH)_2$) and talc ($Mg_3Si_4O_{10}(OH)_2$) as presented in Figure 2.2.

In nature, rocks at or near the surface are rarely free from any hydration, and indeed rocks that are totally hydrated are common. However, many occurrences fall between these extremes. Figure 2.3 shows this continuum of compositions and demonstrates the wide range of possible mineralogical mixtures in naturally occurring ultramafic rocks.

The compositional field of ultramafic rocks forms a prism (Figure 2.3), but trying to plot and compare compositions is difficult

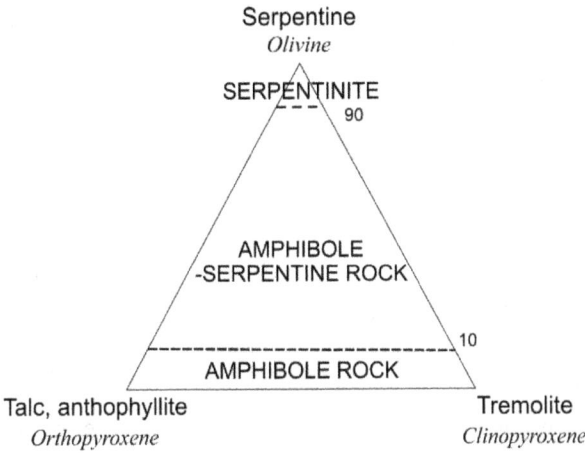

Figure 2.2   Classification of hydrated ultramafic rocks. The parent minerals are provided in italics (from Styles *et al.*, 2014).

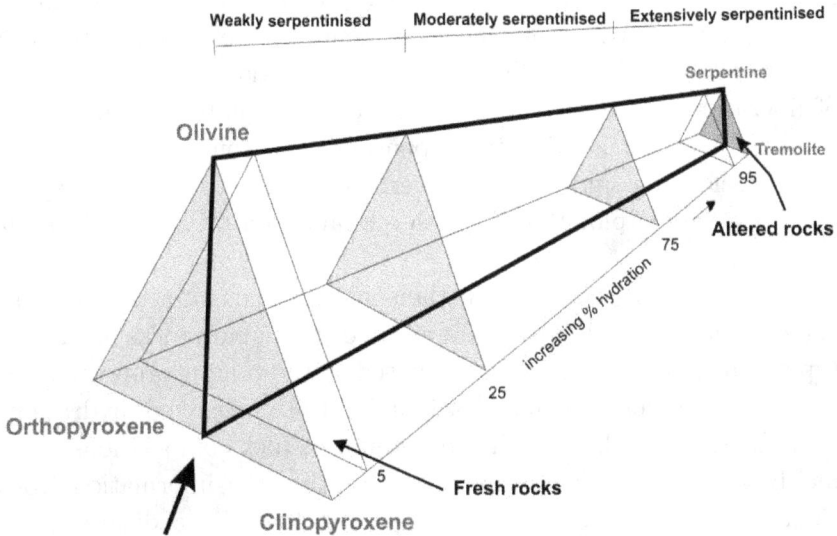

Figure 2.3   The range of mineralogical compositions of naturally occurring ultramafic rocks, showing the plane through the apex used for classification (from Styles *et al.*, 2014).

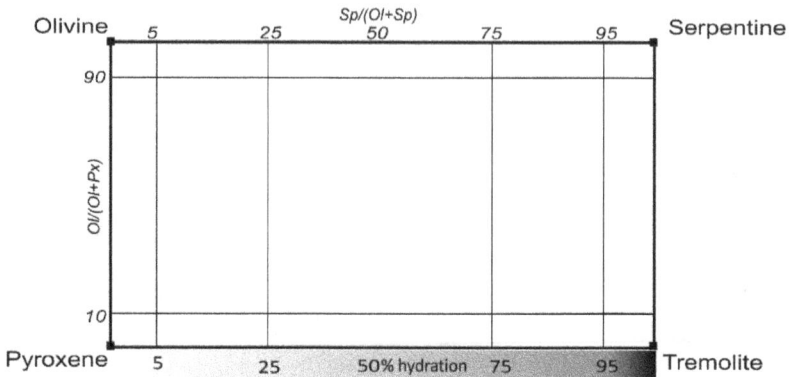

Figure 2.4 Classification diagram for ultramafic rocks based on the proportions of constituent minerals (from Styles *et al.*, 2014).

in a 3D space. However, it is inferred from previous studies of mineral carbonation and from results reported by Styles *et al.* (2014) that (i) orthopyroxene and clinopyroxene behave similarly but are significantly inferior to olivine in terms of cation content and ease of liberation and that (ii) various types of amphiboles also behave similarly but are inferior to serpentine. The compositional prism can therefore be simplified to a vertical section through the apex, to give a simple plot that is most relevant to a CCSM application (Figure 2.4).

The vertical section can then be divided into composition classification fields (Figure 2.4) and used to compare rocks of various types. The plot of compositions is based on the volume proportions of constituent minerals, being closely allied to the degree of hydration. This diagram emphasises the continuum of rock compositions commonly encountered in nature. Incorporating the information from Figures 2.1 and 2.2, the minerals at the top of the diagram, i.e. olivine and serpentine have higher Mg contents, while those at the bottom contain more Si but less Mg, making them less useful for CCSM.

The nature of ultramafic rocks is complex and most natural occurrences, large enough to be a resource for CCSM, will exhibit compositional heterogeneity, manifested by a mixture of the

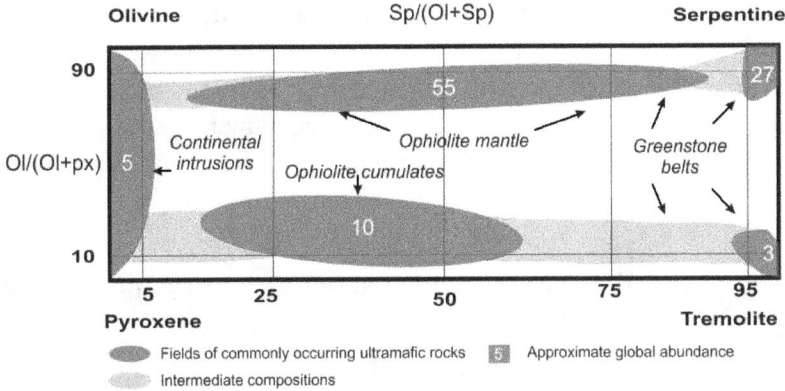

Figure 2.5 The range of compositions of ultramafic rocks and an indication of the proportion (area %) of the global abundance (from Styles *et al.*, 2014; Bide *et al.*, 2014).

anhydrous primary minerals and secondary hydrous minerals. This compositional heterogeneity will have an inevitable effect on their use as a source material. Further, the different types of rocks are not present in equal abundances worldwide; some are much commoner than others. A rough estimate of the global proportions of the various possible types is shown in Figure 2.5. The detail of the geological setting of the various types of rock is not important to this account of the basic principles. The figure shows that one of the most important settings particularly for ultramafic rock is ophiolites. These are fragments of the ocean crust and particularly sub-oceanic mantle that have been pushed up on to the margin of the continent during tectonic movements.

2.4 Mafic Rocks

Mafic rocks are broadly similar to ultramafic rocks but contain a lower proportion of mafic minerals, by definition $< 90\%$ but for the most rocks this is often as low as 50%. The commonest rock by far is the volcanic rock basalt that is found in many places, including the well-known examples of Hawaii and Iceland. A less common, deep seated intrusive equivalent of basalt is known as gabbro. Apart from Mg, Fe silicates, the mafic rocks contain a significant amount of plagioclase feldspar [$(CaNa)Al_2Si_2O_8$] that contains Ca, a potentially useful cation for CCSM. However, the ease of liberation of Ca from plagioclases is much lower than Mg or Fe from mafic minerals. Upon hydration, the mafic rocks, similar to ultramafic rocks, often become enriched in hydrous equivalents of mafic minerals in addition to clay minerals, resulting from alteration of plagioclase, in particular.

2.5 Natural Weathering and Carbonation of Mafic and Ultramafic Rocks

Mafic and ultramafic rocks are far from equilibrium with the atmosphere and surface water, and thus they are very susceptible to weathering. The weathering reaction is exothermic and occurs spontaneously at near ambient pressure and temperature. The natural weathering of mafic and ultramafic rocks has been taking place for billions of years; the rocks are readily hydrated and/or carbonated and as such they play an important part in the CO_2 balance of the earth. The rate and efficiency of mineral breakdown greatly depends on the climatic conditions and rock affected. The mobilised cations pass into solution and either drop out as new phases upon reaching saturation (carbonates, oxides/oxyhydroxides or hydrated silicates) or make their way to the sea where they may react with dissolved CO_2 and are incorporated into animal parts such as shells or ultimately crystallise out as limestone. The latter process is however very slow, taking thousands and possibly millions of years.

2.5.1 *Ultramafic Rocks*

Weathering and carbonation of ultramafic rocks are commonly spatially associated and as such they can be treated as a natural analogue of CCSM, representing both rock dissolution and subsequent CO_2 sequestration. Some of the best examples of such analogues are described from regions where the ultramafic rocks of ophiolites are well exposed on the surface. In such settings, the rocks react with atmospheric CO_2 and/or CO_2 dissolved in ground water. An extreme example of this has been described from the Barzaman Formation in the UAE (Lacinska *et al.*, 2014), where parts of the Hajar

Mountains near the Indian Ocean Coast are composed almost entirely of partially serpentinised ultramafic, mantle peridotite, a component of the Oman–UAE ophiolite. The mantle rocks were emplaced on to the edge of the continent during tectonic movements and uplift, forming the mountains about 5 million years ago. This, in turn, caused rapid erosion and the resulting sand and gravel formed large alluvial fans spreading out towards the Arabian Gulf. Since, the constituent minerals, i.e. serpentine and particularly olivine, are very unstable in the near surface environment, they readily react with the CO_2 dissolved in the groundwater, break up and furnish the solutions with divalent cations that ultimately are bound with CO_2 and Ca from the groundwater to form Mg, Ca carbonate, dolomite (Figure 2.6).

The extent of carbonation of the finer grained sand matrix is generally greater than the larger pebbles. It has been estimated that this natural carbonation at near ambient pressure and temperature ($< 50°C$) close to the earth's surface has locked up around 150 BT of CO_2, equivalent to about 4 years of global emissions at current rates.

Figure 2.6 Carbonated conglomerate (gravel) where the matrix and most of the clasts have been converted from silicate to carbonate minerals in the near surface environment, Barzaman Formation, UAE.

Elsewhere, in the region of Ligurian Ophiolite (Italy), the serpentinites were extensively transformed to magnesite, with subordinated dolomite, also at shallow crustal levels (Malentrata magnesite deposit) (Boschi *et al.*, 2009). This area is characterised by intense hydrothermal activity related to prolonged Pliocene to present magmatism. As a result, the serpentinites were carbonated at elevated temperatures, but probably not exceeding *ca.* 200°C. Importantly, this study shows that pervasive and cyclic hydraulic fracturing maintained a high structural permeability during the whole hydrothermal event and created conduits for the input and output fluids, thereby shedding light onto the feasibility of *in-situ* CCSM in particular.

2.5.2 *Basalt as an Example of a Mafic Rock*

During natural weathering of basalt, the constituent minerals, dominated by pyroxenes and feldspars breakdown and the cations released either drop out of solution as secondary phases or combine with CO_2 from the atmosphere to form bicarbonates and ultimately flow to the sea. Basaltic rocks are generally widely distributed around the globe and some of the largest bodies, resulting from giant volcanic eruptions, are called basalt traps or flood basalts, e.g. Deccan Traps, Siberian Traps or Columbia River Basalts. The emplacement and weathering of such large basaltic provinces play an important role in geochemical and climatic changes on earth and produce a net CO_2 sink on geological time scale (Dessert *et al.*, 2001). For example, the Deccan Traps in NE India that were extruded around 65 Ma and occupy approximately 3 million km^3 are now deeply weathered and two-thirds of the body has been eroded away (Dessert *et al.*, 2001). This weathering has captured around 0.39×10^{12} mol/yr of CO_2 from the atmosphere equivalent to *ca.* 1.8×10^7 US tonnes/year, making *ca.* 12% of the total 1.5×10^8 US tonnes/year CO_2 consumption in basalts globally (Dessert *et al.*, 2003).

2.6 Enhanced Weathering

The previous section shows that natural weathering can result in extensive carbonation of ultramafic rocks and impact the global carbon cycle but the process can be very slow, possibly taking thousands of years. Schuiling and Krijgsman (2006) have proposed using the inherent instability and natural carbonation of olivine to capture CO_2. Olivine could be mined or re-used from previous operations and spread thinly over the ground, possibly accompanying fertiliser application, to undergo natural carbonation. The time suggested for carbonation was around 30 years. Other studies show that in warm wet soils, the actions of fungi and bacteria, around the roots of plants and digestion in the gut of lugworms, can increase the carbonation rate by two orders of magnitude (Schuiling and de Boer, 2013). Schuiling and de Boer (2013) further propose that olivine deposited in the sea, in areas of strong currents, will undergo much more rapid carbonation due to the continuous agitation and attrition that constantly removes any surface reacted layer. Laboratory simulation showed significant reaction in a period of a few weeks. This could be a further option for a low energy penalty method of carbonation.

The role of plants and processes in the soil has been described by Manning and Renforth (2013). They stated that on a global scale, the amount of CO_2 that passes through the plant/soil system in 7 years is equivalent to all the CO_2 in the atmosphere. They also emphasise that using knowledge of these processes provides an opportunity to design artificial soils and land use systems to maximise CO_2 uptake and carbonate precipitation. The main costs and energy used and the implication of trying to implement enhanced weathering in the UK have been described by Renforth (Renforth, 2012). He reported on a model based on adding powdered rock to 50% of the arable land used in the UK. Several factors have to be considered in cost estimation of such models, including mining the feed rock, transport and particularly the fine grinding that is necessary for both wide spreading and increasing surface area to speed reaction rates. In the UK setting, the cost of enhanced weathering of mafic rocks is estimated at £44–361/tCO_2 and for ultramafic rock £15–77.2/tCO_2.

In situations, where these types of rocks have been used for industrial process, the natural weathering rate is greatly increased, primarily due to enhanced surface area resulting from rock crushing to extract valuable minerals. The fine-grained waste, compared to large masses of rock encountered in weathering of natural rock at outcrop, has the potential for much faster carbonation. Of particular relevance are the waste tailings from nickel and asbestos mining that are largely composed of crushed serpentinite.

Studies of the asbestos mine tailings at Clinton Creek in western Canada (Wilson *et al.*, 2009) showed that $164,000 \pm 10\%$ tonnes of CO_2 had been bound and converted to hydrous Mg carbonates in a period of less than 30 years. Various types of reaction had taken place and the formation of a reacted crust may have created a barrier and hindered further carbonation. This was a natural process with no intervention and no energy or cost penalty, and yet the extent of carbonation was at least an order of magnitude greater than that expected for typical weathering. In a situation like this, minor intervention, such as turning over the tailings to expose new reactive material, might significantly increase the extent of carbonation.

A further example of carbonation of serpentinite mine tailings is given by Harrison *et al.* (2013) from the huge Mount Keith nickel mine in Western Australia. Here, brucite $[Mg(OH)_2]$ is present as a small amount of the ~ 11 M tonnes of tailings and natural carbonation is taking place utilising atmospheric CO_2. Experiments show that the limiting factor for carbonation is concentration of CO_2 in the fluid. Mixing the tailings with CO_2 at a higher partial pressure greatly increases the rate and that even flue gas from the mine power plant with $\sim 17\%$ CO_2 and/or utilisation of microbial medium could accelerate carbonation and possibly even offset the emissions from the mine (Harrison *et al.*, 2013; Power *et al.*, 2011).

2.7 Reactivity of Minerals and Rocks

Ultramafic rocks are composed of a limited range of minerals. It is clear from the progress of mineral alteration during weathering that they do not at all behave in the same way and some react faster than others. The *ex-situ* method will involve various industrial processes to liberate the cations from the silicate minerals and make them available for carbonation. It is therefore critical to know how quickly and easily the different minerals release their cations. If *ex-situ* CCSM is to be implemented on a large scale, enormous volumes of rock will have to be processed and hence reaction times must be fast, probably a few hours or ideally much less. A range of processing techniques have been tested to accelerate the release of cations from the silicates, including: acid and alkali leaching (Lackner *et al.*, 1995; Park and Fan, 2004; Wang and Maroto-Valer, 2011; Teir *et al.*, 2007), heating (particularly for serpentine) (O'Connor *et al.*, 2001; McKelvy *et al.*, 2004) and ultrafine grinding (Highfield *et al.*, 2012; Gerdemann *et al.*, 2003).

Various experiments have been carried out to investigate the properties affecting the release of cations, here referred to as reactivity. A good example of such is the series of acid leaching experiments described by Styles *et al.* (2014), the results of which are plotted in Figure 2.7. They used a set of rocks, each composed essentially of only one of the minerals that constitute most ultramafic rocks. The results show that lizardite serpentine has very good reactivity with

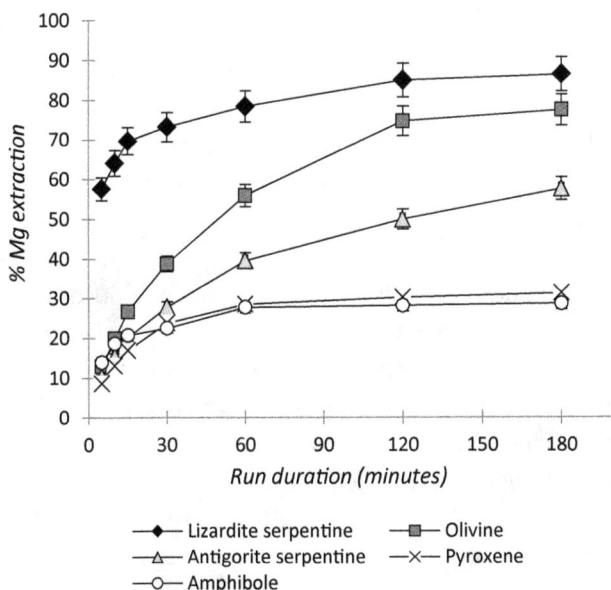

Figure 2.7 The reactivity of mafic minerals during acid leaching based on the amount of Mg extracted as a function of time (after Styles *et al.*, 2014).

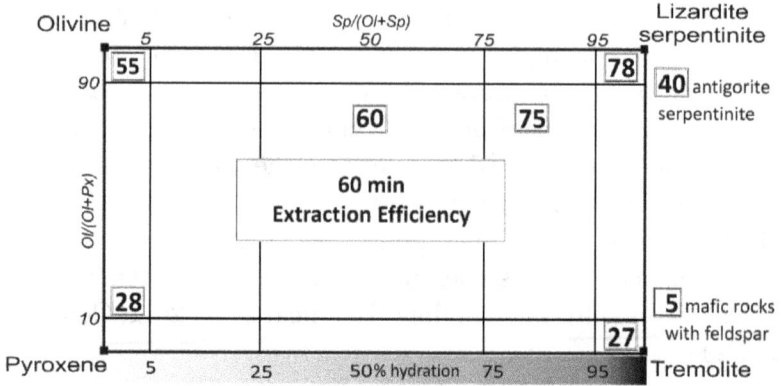

Figure 2.8 Extraction efficiencies of the tested ultramafic rocks in terms of composition — reaction time 60 min (after Styles *et al.*, 2014).

around 80% Mg released within 1 h, compared with pyroxene and amphibole, with only 30%, while antigorite serpentine and olivine showed intermediate values.

These results show that choosing the most reactive rock will have a huge effect on the efficiency and viability of the CCSM technology. Millions of tonnes of feed material will be needed and considerable savings can be made on the amount of rock that needs to be mined, transported and processed if the most reactive are chosen. Using a low reactive rock will increase these amounts by a factor of three. It is probable that high efficiency will be essential to make the *ex-situ* method viable.

These plots show clearly that, considering a 60 min leach time, rocks rich in lizardite serpentine are best. Further, the extraction efficiency decreases significantly as the proportion of olivine increases. Antigorite serpentine and particularly pyroxene and amphibole having efficiencies much less than 50% are probably unacceptable as the cost and energy expended in moving non-productive material would be prohibitive. Mafic rocks that only release 5% of cations are not suited for this process as shown in Figure 2.8 (Styles *et al.*, 2014).

The experiments described above relate solely to activation by acid leaching but it is highly likely that they also have relevance to other chemical and/or thermal activation mechanisms. The purpose of all activation processes is to enhance the speed and efficiency of liberation of cations, particularly Mg^{2+}, from the silicate mineral structure, in simple terms to break the bonds that hold the cations in place within the crystal. The ease with which the bonds can be broken depends, amongst other factors such as P, T, type of solution, etc., on their strength and configuration within the crystal lattice and hence on the amount of energy required to release the cations. Consequently, one of the main factors affecting the ease of cation release is the type of silicate structure. This explains the large reactivity difference between serpentine minerals versus amphiboles and pyroxenes. Serpentines belong to sheet silicates with each sheet containing Mg and Si-hosting layers. The sheets are linked by weak hydrogen bonds that, on the contact with leaching medium, are most prone to destruction. In contrast, the amphiboles and pyroxenes have the silicate groups arranged in chains and the cations are held by much stronger bonds, being therefore more difficult to liberate. Amongst the serpentine mineral group, including lizardite, antigorite and chrysotile, there are small differences in the cation distribution and sheet arrangements and hence the strength of the bonds in their structure, all impacting their reactivity greatly and explaining the significantly higher extraction efficiency of lizardite over antigorite (Lacinska *et al.*, 2016).

Mineral dissolution proceeds via the transport of aqueous reactants to and from the mineral's surface (Schott *et al.*, 2009) leading to permanent alteration of material structure and composition. Chemically, it is a proton–metal exchange with the number of protons exchanged proportional to the valency of the metal (Ludwig and Casey, 1996), e.g. each Mg^{2+} in serpentine would be exchanged with $2H^+$ (Luce *et al.*, 1972; Siever and Woodford, 1979).

The destruction of the slowest breaking metal–oxygen bond constitutes the rate-limiting step during the dissolution of multi-oxide minerals, such as Mg silicates (Oelkers and Schott, 2001; Schott *et al.*, 2009). At low pH persisting during acid leaching, the ease of destruction of metal–oxygen bond decreases with increasing metal valency, e.g. Na–oxygen bond will break more easily than the Si–O bond. Oelkers (2001) presented a comprehensive study on the

		Mineral or solid							
Reaction	Alkali-Feldspar	Anorthite	Muscovite	Kaolinite	Enstatite	Wollas-tonite	Forsterite	Basaltic Glass
Alkali metal -H exchange	Step 1	↓	Step 1			↓		Step 1
Ca-H exchange reaction		Step 1				Step 1	↓	Step 2
Mg-H exchange reaction	↓				Step 1		Mineral Destroyed	Step 3
Tetrehedral Al-H exchange reaction	Step 2	Mineral Destroyed	Step 2					Step 4
Octahedral Al-H exchange reaction			Step 3	Step 1				
Breaking Si-O bonds	Mineral Destroyed		Mineral Destroyed	Mineral Destroyed	Mineral Destroyed	Mineral Destroyed		Solid Destroyed

Decreasing M-O breaking Rate

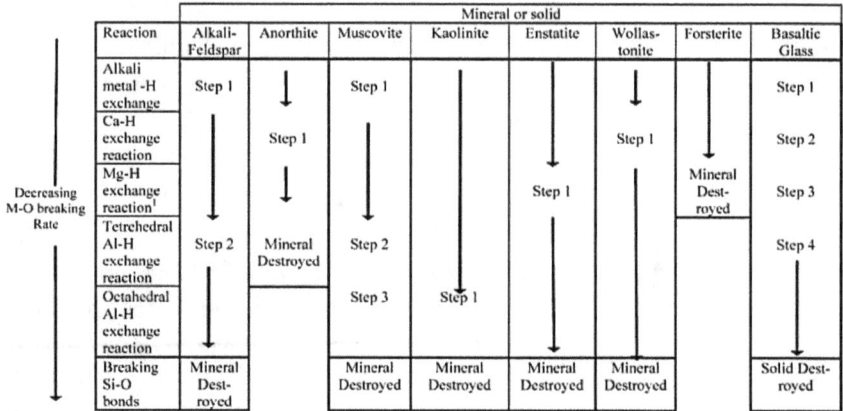

Figure 2.9 Overview of the dissolution mechanisms of various minerals and basaltic glass (Oelkers, 2001).

dissolution kinetics of multi-oxides and summarised the dissolution mechanisms of selected minerals and basaltic glasses under acidic conditions (Figure 2.9). They have concluded that dissolution usually proceeds through several steps, with the "detachment" of alkali metals occurring initially, followed by Ca, Mg and Al.

At low pH, the siloxane (Si–O–Si) bonds are relatively strong (Westrich *et al.*, 1993; Oelkers, 2001) and Si release is relatively slow compared to other metals (Oelkers *et al.*, 2009).

In the case of thermal activation, the mineral structure destruction proceeds via the removal of structurally bound hydroxyls from the solid and their liberation as water vapour. Thermal activation for CCSM is most commonly performed on serpentine minerals and the dehydroxylation temperature varies for different types of serpentine, with the highest $> 700°C$ reported for antigorite and temperatures in the range of *ca.* $650-720°C$ for chrysotile and lizardite (Viti, 2010). Following the dehydroxylation, the serpentine structure becomes largely disordered (poorly crystalline) with Mg^{2+} readily available for subsequent dissolution and/or carbonation. It must be noted that thermal activation of serpentine minerals at temperatures higher than those stated above may result in crystallisation of anhydrous Mg silicates, such as olivine and pyroxene, both having stronger intra-crystalline bonds than serpentines and therefore likely producing an undesirable effect and decreasing the dissolution efficiency. A comprehensive review on dehydroxylation of serpentine minerals with implications for mineral carbonation (Dlugogorski and Balucan, 2014) clearly demonstrates that activation processes must avoid recrystallisation of disordered phases to anhydrous Mg silicates, and minimise the partial pressure of water vapour that engenders reverse reaction.

The reactivity of the minerals is likely to exert a controlling influence on the rock types chosen as feed material for the *ex-situ* process. The *in-situ* method has much less demanding requirements for the speed of reaction but obviously the more reactive minerals will create a better site for carbonation. Little is known about the reactivity of pyroxenes and amphiboles over long periods of time but by analogy to rocks such as the Barzaman Formation described in the previous section, it appears that pyroxene can survive for thousands of years, while all the serpentine and olivine are completely carbonated.

2.8 Global Resources

Ultramafic rocks are widespread across the globe and are suitable for both *in-situ* and *ex-situ* CCSM. For both methods, the CO_2 emitters should be ideally located fairly close to the rock resources and for the *in-situ* method this is almost essential. However, for the *ex-situ* method, it could be possible to transport the rock resources over considerable distances to the emitter sites providing that costs are kept low. This is only likely to be possible where resources and emitters are located close to the coast and cheap transport by ship can be employed. The feasibility of this is shown by the global scale shipping of billions of tonnes of coal and iron ore.

The on-land resources of ultramafic rocks have been estimated by Bide *et al.* (2014) (Figure 2.10). They applied the criteria on mineral reactivity described in Section 2.7 and included only serpentine and olivine-rich rocks. If these rocks were mined to a depth of 100 m only, the potential resources have been estimated at around

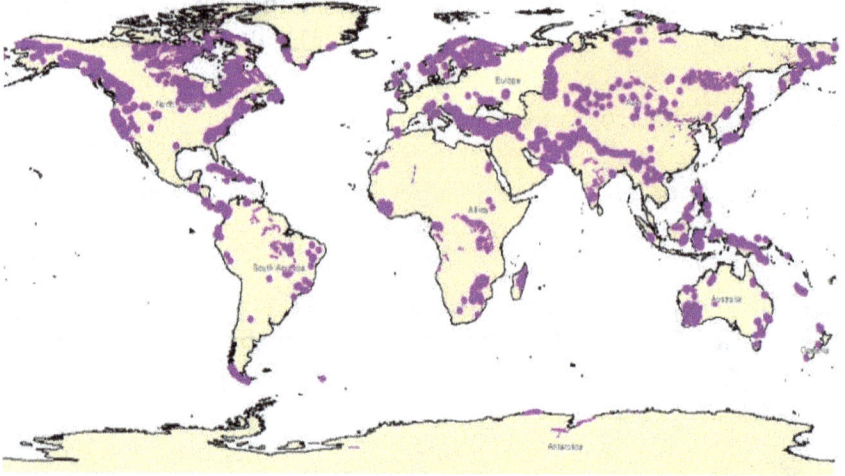

Figure 2.10   Global distribution of ultramafic rocks. Point sources are to show locations only and are not sized proportional to the amount of resources (from Bide *et al.*, 2014).

90 trillion tonnes. Figure 2.10 shows the global distribution of ultramafic rocks. It is clear that many possible resources are located too far from emitters or methods of cheap transport and are simply not feasible for use. However, if only 10% of the worldwide resources were useable, it could mineralise current global CO_2 emissions for 70 years.

The possible locations for *in-situ* CCSM are extensive because volcanic rocks of basaltic composition are also considered as a strong sink for CO_2. Basalt sequences often consist of porous layers of "ash" between layers of impermeable lava that act as a cap rock with properties similar to the situation for conventional underground CCS. The porous ash is likely to react with injected CO_2 and gradually convert to carbonate with a simultaneous decrease in porosity and also lessening the likelihood of leakage as the CO_2 is locked in a stable mineral form. Basalts are very widespread but to be suitable for CO_2 injection, they need to be of considerable thickness and ideally at a depth of $> 1\,\text{km}$. This thickness of basalt is particularly found in flood basalts and large volcanic provinces, such as Siberian Traps or Deccan Traps mentioned before. The total area of these provinces has not been estimated but the Columbia River Province in north-west United States alone has an area of 160,000 km^2. This is around half the area of the ultramafic rocks discussed previously, suggesting that the total resource is orders of magnitude greater than this.

A pilot project investigating CO_2 injection into the Grande Ronde Basalt formation situated within the larger Columbia River Province was undertaken by the Big Sky Carbon Sequestration Partnership. The injection of CO_2 began in July 2013 and 977 metric tonnes of compressed CO_2 was pumped into basalt breccia, in the Wallula pilot borehole, at depth interval of 828 and 887 m (McGrail et al., 2014). Cold CO_2 was heated and pressurised on site before entering the injection well as a supercritical fluid (idem.). Fluid samples collected from reservoir depth showed elevated concentrations of Ca, Mg, Fe and Mn, suggesting a rapid reaction of the injected CO_2 with the reservoir basalt (idem.).

Another example of in-situ CO_2 injection is in Iceland, i.e. the CarbFix Project. Iceland is almost completely made out of basalt. The project encompasses a number of studies on the field-scale injection of CO_2-charged waters into basaltic rocks, along with laboratory-based experiments, large-scale plug flow experiments, a study of natural CO_2 water as a natural analogue and state-of-art geochemical modelling (Alfredsson et al., 2008). The injection site is located in SW Iceland, in the vicinity of the Hellisheidi geothermal power plant. The CO_2 injected into basalts originates from magmatic processes and is the by-product of geothermal energy production. It is reported that more than 80% of CO_2 injected into the CarbFix injection site was carbonated within a year at 20–50°C and 500–800 m depth (Gislason and Oelkers, 2014), providing therefore a high scope for future in-situ CCSM ventures.

There is a further enormous area of basalt that lies beneath the oceans under a thin veneer of sediment (Figure 2.11). However, to utilise the sea-floor formations, the CO_2 captured from industrial sites would have to be transported to the disposal site. Apart from the actual ocean ridges, at the centre of the areas shown in red,

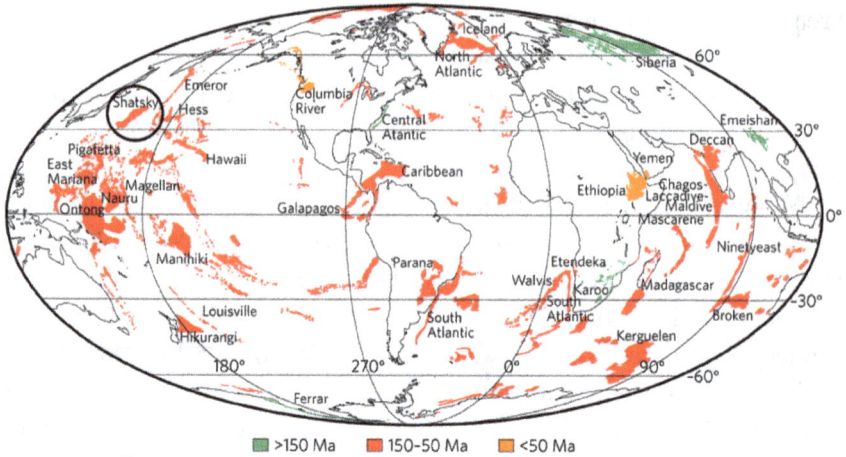

Figure 2.11   Global distribution of large basalt provinces (Coffin and Eldholm, 1994).

the disposal sites will be in very deep water making the CCSM potentially technologically challenging.

The distribution of potential disposal areas clearly shows that location of the rocks in the vicinity of the emitters, or the cost and logistics of transporting CO_2 are a much bigger issue than the volume of rocks available as a repository. The potential for disposal of CO_2 in the ocean ridges has been estimated at around 10 terra tonnes, being sufficient to sequester global emissions for hundreds of years (Slagle and Goldberg, 2011).

2.9 Conclusion

The ultimate objective of CCSM processing is to sequester CO_2 in stable insoluble carbonates, and in practice only Mg and Ca cations are naturally available in quantities that are sufficient to achieve industrial scale carbonation. Accordingly, this defines the type of rocks that are potentially suitable as a CCSM feedstock as those low in silica and high in Mg or Mg + Ca, commonly referred to as ultramafic and mafic rocks. The ultramafic rocks are most suitable as they contain the highest, up to 30 wt.%, proportion of divalent cations. These rocks consist of a number of different minerals, including olivine and pyroxene and their hydrated equivalents serpentine and amphibole. The different types of rocks do not occur in equal amounts, and on a global scale rocks rich in serpentine and olivine are prevalent.

The natural weathering of mafic and ultramafic rocks has taken place over millions of years and the cations released have played an important role in carbon cycle, binding and sequestering CO_2 from the atmosphere as bicarbonate in solution or as solid carbonate minerals. In extreme cases, this has resulted in silicate rocks undergoing complete transformation to carbonate.

Natural weathering can be enhanced in several ways to increase the extent and rate of alteration and carbonate formation. This can be exemplified by thin layer spreading of fine ground rocks on the surface, particularly in warm humid areas characterised by enhanced chemical weathering and activity of plants and invertebrates in the soil; both can greatly increase the rate and extent of carbonation. The deposition of olivine in the sea in areas of high turbulence to increase attrition and dissolution rate has also been proposed. Further, the ongoing natural carbonation of fine grained, serpentine-rich tailings from asbestos and nickel mining could be increased with minimal intervention, for example by agitation, increased water flow to expose more surfaces to alteration and/or adding CO_2 at a higher partial pressure, such as flue gas. As such, these tailing could represent an ideal scenario to pilot *ex-situ* CCSM.

Ex-situ CCSM will require rapid reaction times for extraction of the divalent cations from the silicate minerals, ideally an hour or less. The extraction efficiency of Mg^{2+} from the minerals constituting ultramafic rocks varies greatly. By way of example, the results of acid leaching experiments at 70°C and ambient pressure show that lizardite serpentine exhibits the highest, 80% Mg extraction efficiency in 1 h, while pyroxene and amphibole only 30%, with antigorite serpentine and olivine showing intermediate values.

On a global scale, there are large resources of ultramafic rock that are suitable for both *ex-situ* and *in-situ* CCSM. If for logistical reasons only 10% of the possible resources can be used, this will still be sufficient to sequester 70 years of current global CO_2 emissions.

Mafic rocks, particularly basalt lavas could be used for *in-situ* CCSM in addition to the ultramafic rocks. The volume of these rocks is enormous, greatly exceeding that of the ultramafic rocks and as such they could sequester global emissions for hundreds of years.

In conclusion, suitable rocks to supply feedstock for CCSM are abundant and will not be a constraint on utilisation of the technology. The mineralisation technology needs to develop to the point where it is viable in economic and energy terms to enable these resources to be used.

References

Alfredsson, H. A., B. S. Hardarson, H. Franzson and S. R. Gislason (2008). CO₂ sequestration in basaltic rock at the Hellisheidi site in SW Iceland: stratigraphy and chemical composition of the rocks at the injection site. *Mineralogical Magazine* 72: 1–5.

Bide, T. P., M. T. Styles and J. Naden (2014). An assessment of global resources of rocks as suitable raw materials for carbon capture and storage by mineralisation. *Applied Earth Science* 123: 179–195.

Boschi, C., A. Dini, L. Dallai, G. Ruggieri and G. Gianelli (2009). Enhanced CO₂-mineral sequestration by cyclic hydraulic fracturing and Si-rich fluid infiltration into serpentinites at Malentrata (Tuscany, Italy). *Chemical Geology* 265: 209–226.

Coffin, M. F. and O. Eldholm (1994). Large igneous provinces — crustal structure, dimensions and external consequences. *Reviews of Geophysics* 32: 1–36.

Dessert, C., B. Dupre, L. M. Francois, J. Schott, J. Gaillardet, G. Chakrapani and S. Bajpai (2001). Erosion of Deccan Traps determined by river geochemistry: Impact on the global climate and the Sr-87/Sr-86 ratio of seawater. *Earth and Planetary Science Letters* 188: 459–474.

Dessert, C., B. Dupre, J. Gaillardet, L. M. Francois and C. J. Allegre (2003). Basalt weathering laws and the impact of basalt weathering on the global carbon cycle. *Chemical Geology* 202: 257–273.

Dlugogorski, B. Z. and R. D. Balucan (2014). Dehydroxylation of serpentine minerals: Implications for mineral carbonation. *Renewable & Sustainable Energy Reviews* 31: 353–367.

Gerdemann, S. J., D. C. Dahlin and W. K. O'Connor (2003). Carbon dioxide sequestration by aqueous mineral carbonation of magnesium silicate minerals. Available from: https://netl.doe.gov/publications/proceedings/03/carbon-seq/PDFs/074.pdf

Gislason, S. R. and E. H. Oelkers (2014). Carbon storage in basalt. *Science* 344: 373–374.

Harrison, A. L., I. M. Power and G. M. Dipple (2013). Accelerated carbonation of brucite in mine tailings for carbon sequestration. *Environmental Science & Technology* 47: 126–134.

Highfield, J., H. Lim, J. Fagerlund and R. Zevenhoven (2012). Mechanochemical processing of serpentine with ammonium salts under ambient conditions for CO_2 mineralization. *Rsc Advances* 2: 6542–6548.

Lacinska, A. M., M. T. Styles and A. R. Farrant (2014). Near surface diagenesis of ophiolite-derived conglomerates if the Barzaman formation in the United Arab Emirates: A natural analogue for permanent CO_2 sequestration via mineral carbonation of ultramafic rocks. In: *Tectonic Evolution of the Oman Mountains* [Rollinson, H. R., M. P. Searle, I. A. Abbasi, A. Al-Lazki and M. H. AlKindi's (eds.)]. Geological Society, London, Special publications, vol. 392, pp. 343–360.

Lacinska, A. M., M. T. Styles, K. Bateman, D. Wagner, M. R. Hall, C. Gowing and P. D. Brown (2016). Acid-dissolution of antigorite, chrysotile and lizardite for ex situ carbon capture and storage by mineralisation. *Chemical Geology* 437: 153–169.

Lackner, K. S., C. H. Wendt, D. P. Butt, E. L. Joyce and D. H. Sharp (1995). Carbon dioxide disposal in carbonate minerals. *Energy* 20: 1153–1170.

Luce, R. W., R. W. Bartlett and G. A. Parks (1972). Dissolution kinetics of magnesium silicates. *Geochimica Et Cosmochimica Acta* 36: 35–50.

Ludwig, C. and W. H. Casey (1996). On the mechanisms of dissolution of bunsenite NiO(s) and other simple oxide minerals. *Journal of Colloid and Interface Science* 178: 176–185.

Manning, D. A. C. and P. Renforth (2013). Passive sequestration of atmospheric CO_2 through coupled plant-mineral reactions in urban soils. *Environmental Science & Technology* 47: 135–141.

Mcgrail, B. P., F. A. Spane, C. R. Amonette, C. R. Thompson and C. F. Brown (2014). Injection and monitoring at the Wallula Basalt Pilot Project. *Energy Procedia* 63: 2939–2948.

Mckelvy, M. J., A. V. G. Chizmeshya, J. Diefenbacher, H. Bearat and G. Wolf (2004). Exploration of the role of heat activation in enhancing serpentine carbon sequestration reactions. *Environmental Science & Technology* 38: 6897–6903.

O'Connor, W. K., D. C. Dahlin, D. N. Nilsen, G. E. Rush, R. P. Walters and P. C. Turner (2000). CO_2 storage in solid form: A study of direct mineral carbonation. DOE/ARC-2000-011. Available from: www.osti.gov/scitech/servlets/purl/896225.

Oelkers, E. H. (2001). General kinetic description of multioxide silicate mineral and glass dissolution. *Geochimica Et Cosmochimica Acta* 65: 3703–3719.

Oelkers, E. H., S. V. Golubev, C. Chairat, O. S. Pokrovsky and J. Schott (2009). The surface chemistry of multi-oxide silicates. *Geochimica Et Cosmochimica Acta* 73: 4617–4634.

Oelkers, E. H. and J. Schott (2001). An experimental study of enstatite dissolution rates as a function of pH, temperature and aqueous Mg and Si concentration, and the mechanism of pyroxene/pyroxenoid dissolution. *Geochimica Et Cosmochimica Acta* 65: 1219–1231.

Park, A. H. A. and L. S. Fan (2004). CO$_2$ mineral sequestration: Physically activated dissolution of serpentine and pH swing process. *Chemical Engineering Science* 59: 5241–5247.

Power, I. M., S. A. Wilson, D. P. Small, G. M. Dipple, W. Wan and G. Southam (2011). Microbially Mediated Mineral Carbonation: Roles of Phototrophy and Heterotrophy. *Environmental Science & Technology* 45: 9061–9068.

Renforth, P. (2012). The potential of enhanced weathering in the UK. *International Journal of Greenhouse Gas Control* 10: 229–243.

Schott, J., O. S. Pokrovsky and E. H. Oelkers (2009). The Link Between Mineral Dissolution/Precipitation Kinetics and Solution Chemistry. In: *Thermodynamics and Kinetics of Water-Rock Interaction* [Oelkers, E. H. and J. Schott, (eds.)].

Schuiling, R. D. and de Boer, P. L. (2013). Six commercially viable ways to remove CO$_2$ from the atmosphere and/or reduce CO$_2$ emissions. *Environmental Sciences Europe* 25: 35, 9.

Schuiling, R. D. and P. Krijgsman (2006). Enhanced weathering: An effective and cheap tool to sequester CO$_2$. *Climatic Change* 74: 349–354.

Siever, R. and N. Woodford (1979). Dissolution kinetics and the weathering of mafic minerals. *Geochimica Et Cosmochimica Acta* 43: 717–724.

Slagle, A. L. and D. S. Goldberg (2011). Evaluation of ocean crustal Sites 1256 and 504 for long-term CO$_2$ sequestration. *Geophysical Research Letters*, 38(L16307): 1–5.

Styles, M. T., A. Sanna, A. M. Lacinska, J. Naden and M. Maroto-Valer (2014). The variation in composition of ultramafic rocks and the effect on their suitability for carbon dioxide sequestration by mineralization following acid leaching. *Greenhouse Gases-Science and Technology* 4: 440–451.

Teir, S., H. Revitzer, S. Eloneva, C.-J. Fogelholm and R. Zevenhoven (2007). Dissolution of natural serpentinite in mineral and organic acids. *International Journal of Mineral Processing* 83: 36–46.

Viti, C. (2010). Serpentine minerals discrimination by thermal analysis. *American Mineralogist* 95: 631–638.

Wang, X. and M. Maroto-Valer (2011). Dissolution of serpentine using recyclable ammonium salts for CO$_2$ mineral carbonation. *Fuel* 90: 1229–1237.

Westrich, H. R., R. T. Cygan, W. H. Casey, C. Zemitis and G. W. Arnold (1993). The dissolution kinetics of mixed-cation orthosilicate minerals. *American Journal of Science* 293: 869–893.

Wilson, S. A., G. M. Dipple, I. M. Power, J. M. Thom, R. G. Anderson, M. Raudsepp, J. E. Gabites and G. Southam (2009). Carbon Dioxide Fixation within Mine Wastes of Ultramafic-Hosted Ore Deposits: Examples from the Clinton Creek and Cassiar Chrysotile Deposits, Canada. *Economic Geology* 104: 95–112.

Chapter 3

CO$_2$ Mineralisation as a Route to Energy-Efficient CO$_2$ Sequestration or Materials with Market Value

Ron Zevenhoven* and Inês S. Romão*,†

*Åbo Akademi University, Turku, Finland
†University of Coimbra, Portugal

3.1 Introduction

The fixation of CO$_2$ in inorganic carbonates has, despite its practically limitless potential and other documented benefits, not developed into mature technology that finds widespread and large-scale application today. For many, carbon capture and storage (CCS) has become synonym to carbon capture and geological storage (CCGS) technology that involves underground storage of purified, pressurised CO$_2$, at economically acceptable distances from where the CO$_2$ is produced. See, for example, the European Commission (EC) directive on CCS from 2009 (EC, 2009). This was certainly motivated by the successful start-up of three large-scale CCS demonstrations employing underground saline aquifers — Sleipner, Snøhvit and In-Salah in Norway and Algeria, in 1996, 2008 and 2004, respectively — together fixing around 3 Mt CO$_2$ annually. As these projects involved separation of CO$_2$ from natural gas, the support from oil/gas processing industry was strong, setting the tone of the International Energy Agency (IEA) implementing agreement on greenhouse gas (GHG), its flagship *International Journal of Greenhouse Gas Control* (IJGGC) and the *Greenhouse Gas Control Technologies* (GHGT) conference series.

At the same time, the power industry and metal/mineral/cement industry sector were facing the problem of having to separate CO_2 from oxygen-rich gases for which the typically used alkanolamine solvents are less suitable, suffering from degradation and leading to significant losses while forming toxic nitrous compounds (Knudsen *et al.*, 2007; Mazari *et al.*, 2015). As a result, as alternatives to post-combustion CO_2 removal, some attention moved to pre-combustion CO_2 removal and oxy-fuel combustion, shifting the separation to either CO_2 removal from gasifier syngas or air separation. Similar to the typical energy penalty of 3–5 GJ/tCO_2 for CO_2 removal from natural gas using amines, the energy penalty of pre-combustion and oxy-fuel gas separation are significant as well. Although gasification and oxy-fuel-based processes using fossil carbon fuels are slowly and undeniably gaining ground, it is still CCGS that determines the CCS agenda, not to mention its image to the public that has more or less discarded the option of on-shore CCGS.

Much of the CCS deployment credit is today given to CCGS involving enhanced oil recovery (EOR): being applied primarily in North America the current annual use of CO_2 for this is around 30 Mt. Unfortunately, extracting fossil carbon eventually gives CO_2 upon use as fuel and indeed, depending on how much CO_2 dissolves in the oil produced, using EOR–CCS gives an increased CO_2 production of 1.3–2.6 t/tCO_2 used for EOR (Armstrong and Styring, 2015). Off-setting this with CCGS using saline aquifers would require significant extra storage capacity, and the current (November 2015) situation of 1.6 Mt CO_2/a CCGS using saline aquifers (three projects) plus 25.8 Mt CO_2/a EOR–CCGS (11 projects) suggests that CCGS is not only a slow-mover under the umbrella of CCS/CCUS technologies, but also is moving in a direction where *de facto* no CO_2 emissions are mitigated. (Besides this, there is mixed reporting as to what is happening with CO_2 once stored underground, considering CO_2-related chemistry and storage site integrity.) A breakthrough is certainly the Boundary Dam project in Canada where, since

October 2014, CO_2 is removed from coal-fired power plant flue gas at a rate of ~3.5 tonnes/day (1 Mt/a), to be used for EOR ~65 km away (IEA GHG, 2015). This first demonstration plant operating on coal-fired power plant flue gas gives much information on the amine-type solvent performance while much ongoing research focuses on solvents that are more stable, less volatile and corrosive and bring a lower regeneration energy (i.e. heat) penalty than the typical 3–5 GJ/tCO_2 (Park *et al.*, 2015). Yet, other separation methods are being developed as alternatives, as for example, membranes which were reported (for CO_2/N_2 separation) to be more energy efficient than amine absorption when CO_2 concentration is >20 vol% (Khalilpour *et al.*, 2015).

Although separating CO_2 from oxygen-containing flue gases and process gases is far from common practice, options of utilising CO_2 while generating revenue is not limited to EOR. Several carbon capture and utilisation (CCU) routes have been suggested where CO_2 is used as a chemical feedstock or solvent. Indeed, market volumes have increased to a current use of ~200 Mt CO_2/a, with urea being the major product (Armstrong and Styring, 2015). Although this is several times the stream of CO_2 used for EOR and dwarfs the CO_2 stored in saline aquifers, it must be noted that for many CCU routes, the duration of "storage" in most CCU products falls short of the (several) 1,000 years aimed at for CCS. Besides this, most CCU methods will have energy and/or hydrogen input requirements, presenting a challenge that brings renewable energy and water into the picture.

Mineral carbonation (MC), (or more generally CO_2 mineralisation) has so far been an odd part of the CCS portfolio, seen by many as a "competitor" for CCGS (and the struggle for funding for that), while many others classify MC as a CCU method. The first part of this chapter summarises the developments, with an emphasis on progress made during the last 5 years. The trends listed above and primarily the lack of deployment of CCGS (ex-EOR) have, nonetheless, resulted in an increasing stream of publications and patented ideas in the field of MC. There is a fair chance that industries that see the processing of alkaline waste streams and by-products, while sequestering CO_2 as a "bonus", will create large-scale markets. As will be addressed in the second part of this paper, energy use and, for a better overall picture, life cycle impact (LCI) performance can eventually bring MC under the attention of the fuel, heat and electricity-producing industries. Most researchers investigating MC do not address or report energy use: the work of those who do is concentrated in the second part of this chapter. In the end, all CCS/CCUS methods should present answers to the questions on how to stop or slow down climate change and global warming resulting from the use of fossil fuels as an energy source.

3.2 CO_2 Mineralisation Technology Development

The natural weathering of rocks is a process that occurs at very low rates in nature and that has been responsible for the uptake of billions of tonnes of CO_2 from the atmosphere throughout times. MC, as a fast (and possibly cost-effective) industrial version of the natural weathering of rocks, was first brought up by Seifritz (1990). The final goal of MC is to trap CO_2 into benign carbonates that will sequester CO_2 for thousands of years. A major advantage comes from the fact that carbonates will remain in the solid form (long-time scales) as they have a lower energy state than CO_2 (Lackner *et al.*, 1995). In fact, stability of carbonates, even under strong acidic conditions, was confirmed by Teir *et al.* (2006b) and later also by Allen and Brent (2010). Both authors concluded that, under all plausible conditions of pH and rainfall, leakage rates of stored CO_2 are insignificant.

To date, research has mainly focused on carbonation of alkaline materials such as natural minerals and industrial wastes. Therefore, CO_2 mineralisation is a more appropriate name that covers all processes that bind CO_2 into a solid inorganic carbonate (Power *et al.*, 2013) while for this chapter MC is appropriate, leaving industrial wastes largely outside the discussion. The potential for CO_2 sequestration via MC is enormous at 10,000–1,000,000 Gt CO_2 (see Figure 3.1) as the amounts of suitable and readily available mineral silicates far exceed requirements for sequestering all anthropogenic CO_2 emissions from the known fossil fuel reserves (Lackner *et al.*, 1995). This makes MC particularly attractive for large point

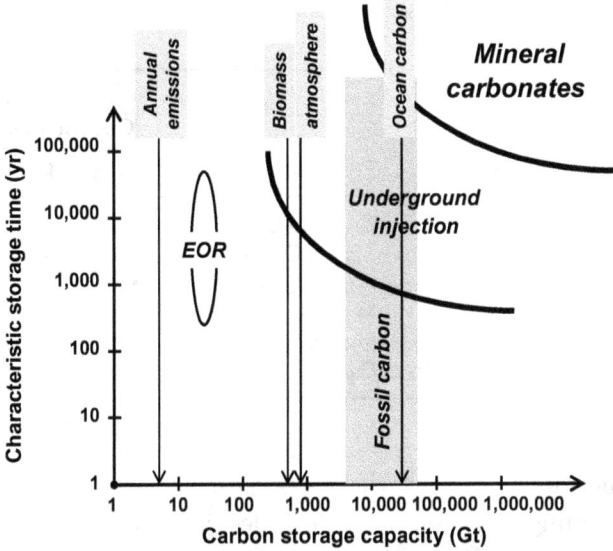

Figure 3.1 Estimated times and storage capacities for the various options for CO_2 storage (taken from Zevenhoven *et al.*, 2006, adapted from Lackner, 2003).

CO_2 sources: power generation, iron and steel making and cement manufacture, for example.

The chemistry behind CO_2 mineralisation is relatively simple: The minerals react with CO_2 (separated from flue gases or directly with the CO_2-containing flue gases) to form a carbonate as shown in the general Reaction (3.1), where M stands for the divalent metal Mg, Ca (or in some cases Fe):

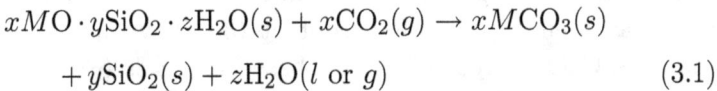

$$xMO \cdot ySiO_2 \cdot zH_2O(s) + xCO_2(g) \rightarrow xMCO_3(s)$$
$$+ ySiO_2(s) + zH_2O(l \text{ or } g) \tag{3.1}$$

Depending on the mineral in reaction (given by values for x, y and z in the reaction), the overall heat output ranges from $+50$ to $+100$ kJ/mol of CO_2 converted. As an example, the carbonation of serpentine — Reaction (3.2) — releases ~ 64 kJ/mol CO_2 fixed.

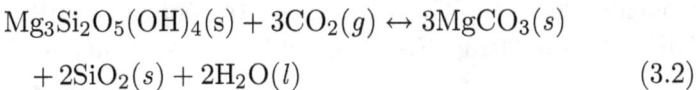

$$Mg_3Si_2O_5(OH)_4(s) + 3CO_2(g) \leftrightarrow 3MgCO_3(s)$$
$$+ 2SiO_2(s) + 2H_2O(l) \tag{3.2}$$

Table 3.1   Chemical composition of minerals suitable for MC. Adapted from Penner *et al.* (2004).

Rock/Mineral Group	Mineral	Formula	Chemical Composition (wt.%)		
			Ca	Fe²⁺	Mg
Feldspar	Anorthite	$CaAl_2Si_2O_8$	10.3	3.1	4.8
Serpentine	Antigorite	$Mg_3Si_2O_5(OH)_4$	< 0.1	2.4	24.6
Pyroxene	Augite	$CaMgSi_2O_6+(Fe,Al)$	15.6	9.6	6.9
Olivine	Fosterite	Mg_2SiO_4	0.1	6.1	27.9
Ultramafic	Talc	$Mg_3Si_4O_{10}(OH)_2$	2.2	9.2	15.7
Ultramafic	Wollastonite	$CaSiO_3$	31.6	0.5	0.3

The exothermic nature of the carbonation reaction hints towards a self-sustaining process. However, despite the simple chemistry, recreating the natural weathering at an industrial time scale is a thorny challenge, mainly due to slow kinetics and thermodynamic limitations (Huijgen and Comans, 2005; Zevenhoven *et al.*, 2008; Torróntegui, 2010; Fagerlund, 2012; Nduagu *et al.*, 2012a; Olajire, 2013).

Natural Mg/Ca silicate minerals (olivines, serpentines and basalts) are cheap and abundant worldwide (see Figure 2.10), and highlighted as good candidates for MC (e.g. Seifritz, 1990; Lackner *et al.*, 1995; Goff and Lackner, 1998; Huijgen and Comans, 2005; Sipilä *et al.*, 2008). Typical contents of naturally occurring ultramafic rocks are presented in Table 3.1. Magnesium-based silicates (olivines, serpentinites) are preferable for large-scale CO_2 sequestration purposes, as their typical MgO content (35–50 wt.%) is much higher than the CaO content of the Ca-based minerals (~10 wt.%), which would require the handling of impractical quantities of calcium-based material resources for carbonation.

3.3   MC Routes

For the last two decades, several reviews on MC (e.g. Huijgen and Comans, 2005; Sipilä *et al.*, 2008; Torróntegui, 2010; Khoo *et al.*, 2011; Olajire, 2013; IEA GHG, 2013; Sanna *et al.*, 2014; Azdarpour *et al.*, 2015) have been reporting advances in MC technology, mainly addressing the challenges imposed by the slow kinetics and thermodynamic limitations imposed by the carbonation process. All the reviews come to the same conclusions: in general MC is too energy intensive, requires extreme operating conditions and the handling of large amounts of water, chemicals and minerals. Furthermore, the technology is not mature and pilot scale demonstration projects are indispensable.

MC is commonly seen as an inferior option towards CCS involving geological storage, mainly due to (1) on the short run is it more energy intensive; although often estimation of energy inputs (and therefore costs) are distorted by wrong assumptions giving an unrealistic view of the economy of the process (Zevenhoven *et al.*, 2008; Zevenhoven and Fagerlund, 2010) and (2) it involves the mining of considerable amounts of minerals and the handling of high quantities of chemicals and water. It is, however, important to keep in mind that (under the right conditions) MC may remove the conventional separation/purification step of CO_2 from the CCS chain (Verduyn *et al.*, 2009). After all, most components of the flue gas will be inert during the carbonation process (requiring only a large gas flow compared to a purified CO_2 stream, which becomes less of a problem when pressurised), while co-binding of SO_2 is an interesting option (Zevenhoven *et al.*, 2012). Still, at this point, it is clear that even if safer, MC will only thrive through integration with large-scale CO_2 producers (Slotte *et al.*, 2013), for reducing net CO_2 emissions and overall energy (primarily heat input), and by producing marketable products. Yet it will be a challenge to deploy MC on a large scale because one significant drawback compared to CCGS is that the CO_2 stays above the ground and is supposed to form a part of a carbon-based materials economy. Mehleri *et al.* (2015) recently presented a techno-economic assessment on a few CCS/CCUS route and conclude MC to be neither cost-effective nor scalable to the volumes needed for substituting CCGS. This can be mirrored against an earlier assessment that mentions MC as a value "green economy" technology to be applied sequentially to CCGS when storage capacity runs out (Styring *et al.*, 2011).

Nonetheless, R&D work continues and MC process routes under development can be divided in two main groups: *in-situ* and *ex-situ* accelerated MC. *In-situ* carbonation MC and the third group of routes, enhanced weathering, are addressed in other chapters of this book volume and won't be further considered.

3.4 *Ex-Situ* Accelerated MC

Due to the complexity of the *ex-situ* mineralisation processes, the categorisation of some studies is somewhat ambiguous. Commonly, the mineralisation routes are classified as direct (Mg/Ca extraction and carbonation take place in one step) and indirect processes (the Mg/Ca extraction and carbonation are sequential and take place in independent steps). Both direct and indirect methods can be further divided into dry and aqueous carbonation routes. A more detailed classification is presented in Figure 3.2.

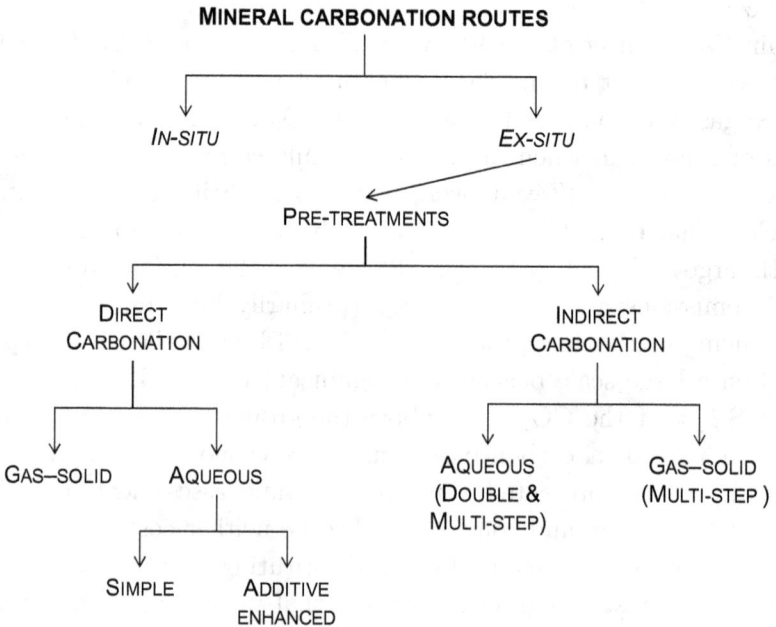

Figure 3.2 MC pathways for naturally occurring minerals.

3.4.1 *Mineral Pre-treatment Options*

For *ex-situ* carbonation, conventional crushing and grinding are obligatory with typical particle sizes ranging from 50 to 250 μm. Activation of the mineral feedstock may be achieved by mechanical (Gerdemann *et al.*, 2007; Baláž *et al.*, 2008; Haug *et al.*, 2010; Turian-icová *et al.*, 2013), thermal (McKelvy *et al.*, 2004; Boerrigter, 2010; Balucan *et al.*, 2013; Balucan and Dlugogorski, 2013; Dlugogorski and Balucan, 2014) and chemical means (Zhang *et al.*, 1996; Maroto-Valer *et al.*, 2005; Nduagu, 2008, 2012b). Typically, mechanical activation is applied to olivines while thermal activation is applied to serpentinites (Torróntegui, 2010). The aim is to go beyond surface area increase and, somehow, disrupt the mineral lattice decreasing the activation energy for any subsequent reactions (O'Connor *et al.*, 2005; Haug *et al.*, 2010). Although mechanically activated feedstock materials are more reactive, it is often argued that the high energy input requirements lead to unacceptable energy penalties (Haug *et al.*, 2010; Huijgen *et al.*, 2007). Besides this, some work was reported on biological acceleration of magnesium silicate rock dissolution and/or carbonation. Indeed activity is seen after time scales of weeks/months (Bundeleva *et al.*, 2014). This new field may have future potential but it won't be considered here. (For calcium silicates more promising results have been reported by Salek *et al.*, 2013.)

As for the heat treatment (at 500–700°C depending on the mineral) is recurrently found to be too energy intensive and thus costly (Huijgen and Comans, 2005; IPCC, 2005; Gerdemann *et al.*, 2007). However, Zevenhoven *et al.* (2008), Zevenhoven and Fagerlund (2010) and later Balucan *et al.* (2013) and Brent *et al.* (2015), argue that cost projections for serpentine carbonation in O'Connor *et al.* (2005) and Gerdemann *et al.* (2007), based on activation of mineral feedstock using electrical energy are incorrect and negatively affects the economy of the process.

3.4.2 *Direct Dry (Gas/Solid) Carbonation*

Direct gas/solid carbonation is the simplest approach of MC that, in theory, allows for recovery of the carbonation heat. The direct conversion of gaseous CO_2 into Mg/Ca solid material was first assessed at the Los Alamos National Laboratory (LANL) (Lackner *et al.*, 1995, 1997) and later in Finland at Helsinki University of Technology (TKK) (Kohlmann and Zevenhoven, 2001; Zevenhoven and Kohlmann, 2002). Both research groups concluded that the direct gaseous carbonation of minerals suffers from low conversion and slow reaction kinetics thus requiring high temperature and pressure. At ~340 bar, Lackner *et al.* (1997) achieved a maximum of 30% carbonation of serpentinised powder. Carbonation of pure serpentinite powder at 200–1,000°C in a pressurised thermogravimetric analyser (PTGA) by Zevenhoven and Kohlmann (2002), showed only minor changes in the mineral's carbonate content even after relatively long reaction time. Only almost a decade later, Kwon (2011) attempted the direct carbonation of olivine. Yet again, the author recognises that the low conversion (27% of olivine carbonation in 3.7 μm size particles) due to thermodynamic limitations is the main drawback to overcome (Kwon *et al.*, 2011). More recently, Larachi *et al.* (2012) attempted the carbonation of pre-treated chrysotile (serpentine) for 1 h between 100 and 220°C at 3.2 MPa total pressure for $CO_2/H_2O/H_2$ mixtures. Despite uptakes as high as 0.7 CO_2 moles per mol of magnesium at 130°C, the pre-treatment methods (e.g. thermal activation of chrysotile at 700°C for 20 min) present a prohibitive energy penalty.

As it became clear that direct mineralisation of Mg/Ca natural silicates is an unviable process, research progressed to multi-step (indirect) gas–solid carbonation routes: these will be later addressed in detail.

(As a side note, dry carbonation of industrial alkaline wastes, which are more reactive than minerals, continues to stir interest. The trend is to use flue gases (avoiding the CO_2 separation step) to carbonate and stabilise waste materials — see Sanna *et al.* (2014)).

3.4.3 Direct Aqueous Carbonation

Direct aqueous carbonation involves three co-existent mechanisms:

(i) dissolution of CO_2 into water towards a mildly acidic solution

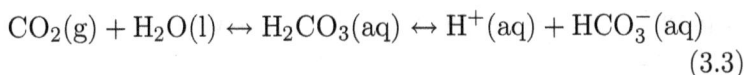

$$CO_2(g) + H_2O(l) \leftrightarrow H_2CO_3(aq) \leftrightarrow H^+(aq) + HCO_3^-(aq)$$
$$(3.3)$$

(ii) to facilitate the dissolution of the Mg/Ca rich material

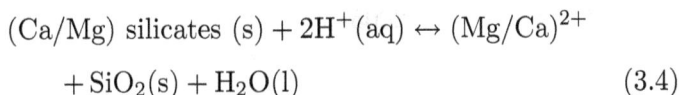

$$(Ca/Mg) \text{ silicates } (s) + 2H^+(aq) \leftrightarrow (Mg/Ca)^{2+}$$
$$+ SiO_2(s) + H_2O(l) \qquad (3.4)$$

(iii) and the subsequent the precipitation of the leached ions as carbonates:

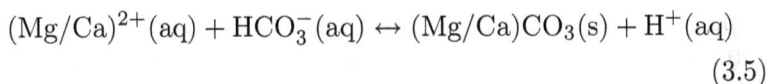

$$(Mg/Ca)^{2+}(aq) + HCO_3^-(aq) \leftrightarrow (Mg/Ca)CO_3(s) + H^+(aq)$$
$$(3.5)$$

The produced carbonates may be in the form of hydrates, for example, hydromagnesite $Mg_5(OH)_2(CO_3)_4 \cdot 4H_2O$. Difficulties arise from the fact that each reaction limits the others: steps (i) and (ii) require an acidic media while step (iii) is favoured at alkaline conditions. Thus, careful optimisation is required and direct carbonation must happen in narrow pH ranges. Temperature optimisation is also tricky as a higher temperature enhances the leaching of magnesium from the mineral lattice but decreases the CO_2 solubility in water.

In general, it is agreed that the main constraint of the aqueous scheme for commercial application is the slow kinetics for the silicate dissolution (Huijgen and Comans, 2005; Sipilä *et al.*, 2008; Fagerlund, 2012; Nduagu, 2012b; Li *et al.*, 2013; Olajire, 2013). Initially the reaction happens at a fast rate (Carey *et al.*, 2003) but, as it progresses, the build-up of a passivating SiO_2 layer on the particle's surface prohibits the further magnesium release from the mineral's matrix (Teir *et al.*, 2007a).

As experimental results showed that CO_2 did not produce enough acidity to accelerate mineral dissolution (O'Connor *et al.*, 2002, 2005), addition of alternative chemicals (organic and inorganic) and feedstock pre-treatments (mechanical and heat activation) became imperative. As so, during the last two decades a wide range of acids and complexing agents were tested: HCl was first studied at the LANL by Lackner *et al.* (1995, 1997) and by Park *et al.* (2003) at the Ohio State University. Studies by Fauth *et al.* (2000) and O'Connor *et al.* (2005) at NETL (former ARC) resulted in an optimised buffered saline solution of 0.64 M $NaHCO_3$ and 1 M NaCl (Gerdemann *et al.*, 2007). The authors reported an 80% conversion of heat pre-treated (615–630°C) serpentine at 155°C and 115 bar. The heat pre-treatment, fine grinding $<70\,\mu$m and the difficulty in recovering the reactants again imply a process too expensive and impracticable for industrial implementation (Huijgen and Comans, 2005; Gerdemann *et al.*, 2007). Brent *et al.* (2015) note that the presence of NaCl in the process solutions presents an equipment corrosion risk and that chloride should be eliminated from the process. Nevertheless, this route was considered the state-of-the-art for many years (it was explicitly mentioned in the IPCC special report on CCS (IPCC, 2005)) and is still a reference and comparison benchmark for many ongoing studies. In fact, recently Eikeland *et al.* (2015) achieved full carbonation of olivine particles in less than 4 h, at 190°C, in a solution of 0.5 M $NaHCO_3$ + 0.5 0.75 M NaCl, but at the cost of extra fine grinding ($< 10\,\mu$m) and 100 bar CO_2 pressure. R&D went on with mineral and organic acids: $KHCO_3$ (Mckelvy, 2006), H_2SO_4, HNO_3, HCOOH, CH_3COOH (Teir *et al.*, 2007a, 2007b), oxalates (Olsen and Donald Rimstidt, 2008, Rimstidt *et al.*, 2012), citrates and EDTA (Park *et al.*, 2003; Krevor and Lackner, 2009, 2011; Rozalen and Huertas, 2013; Bobicki *et al.*, 2014, 2015). So far, no significant breakthroughs were achieved, especially for the organic additives.

Finding an additive to promote dissolution of the silicate mineral, under a pH favourable to carbonates formation, would be a significant step forward for direct aqueous carbonation routes (Krevor and Lackner, 2009). Seeking to overcome the limiting nature of the three co-existent carbonation mechanisms, Blencoe *et al.* (2004), proposed the dissolution of calcium silicates in an alkaline media. The long reaction times ($\sim$72 h) and large amounts of NaOH needed (50–80 wt.%) were considered to be unaffordable for industrial application (Huijgen and Comans, 2005). Still, years later, the same authors patented an identical process for application to a wider range of metal silicates (Blencoe *et al.*, 2012) yet, once more, without any report on the $Mg(OH)_2$ production and carbonation efficiencies. In a recent study, Madeddu *et al.* (2014) obtained an 80 wt.% conversion of dunite (ultramafic rock, mainly olivine) into $Mg(OH)_2$, using a 40% NaOH solution at 180°C (pressure not mentioned) for 24 h. The author also emphasises that producing the NaOH needed to process 1 kg of dunite would produce 13 kg of CO_2, while the capture potential of the processed material is of $\sim$0.53 kg CO_2. Only after 26 recycling loops of NaOH, the process would become carbon neutral. These results utterly discourage the alkaline digestion route.

In the US, Gadikota and co-workers are re-considering the NETL (former ARC) route mentioned above, focusing on the effect of NaCl and bicarbonate compounds in the solutions and the reaction kinetics and MC mechanisms in general, for olivine (Gadikota *et al.*, 2014a). While confirming many earlier findings, the effect of (1 M) NaCl, assumed to be weakly chelating, was found insignificant, and $NaHCO_3$ (used as pH buffer) was found to influence the precipitation of iron oxides. A comparison of freshly ground and aged olivine and antigorite reactivity showed different yields, while also the presence of fine particles ($<5\,\mu$m) can result in overestimated reaction rates (Gadikota *et al.*, 2014b). The group also studied the morphology changes during MC, which can be especially beneficial when processing asbestos-containing material (ACM). A comparison was made with serpentinite, chrysotile and olivine, using sodium oxalate and acetate as ligands. The carbonation of AMC drastically changed its fibrous structure, with a significant dissolution of the ACMs calcium content (38 wt.%), and 0.1 M oxalate was found to be effective in enhancing the carbonation rate (Gadikota *et al.*, 2014c).

Researchers at the Swiss Federal Institute of Technology in Zürich (ETH), focused on the detailed understanding of the direct aqueous

MC chemistry and did an extensive work on olivine dissolution (Hänchen *et al.*, 2008; Prigiobbe *et al.*, 2009a, 2009b) at different temperatures, CO_2 pressure and salinity. The authors use a population balance approach to develop a model to describe the dissolution of olivine. (Hänchen *et al.*, 2007, 2008), also providing insight on Mg-carbonate stability under the given conditions. Once more the authors conclude that only high temperature (150°C) and pressure (100 bar) allow for rates convenient for industrial operation. In the attempt of skipping the costly step of CO_2 purification, R&D at the ETH progressed on routes that make direct use of flue gases. Werner *et al.* (2013) performed single-step carbonation experiments in stirred reactors with gas-dip tubes, with thermally pre-activated serpentine. The authors found that, once the experiments reached equilibrium conditions, the extent of the carbonation reaction did not exceed 20%. This way, they (Hariharan *et al.*, 2013, Werner *et al.*, 2014) initiated dissolution studies far from equilibrium conditions. With a mandatory thermal pre-activation of rock material, in mildly acidic media, moderate partial pressure of CO_2 (2 bar of CO_2) and temperatures of 120°C, in flow-through operation conditions, and as a result 83% of the magnesium was dissolved within 100 min. Based on those experimental results, (Hariharan *et al.*, 2014) developed a kinetic model, for the heat-treated lizardite. More recent work with heat-treated lizardite has progressed to a two-step process, with two reactors operating at different temperatures (30°C and 90°C), in combination with a PCO_2 swing. Significantly higher Mg extraction ($< 30\% \rightarrow \sim 50\%$) and carbonation efficiencies ($< 20\% \rightarrow\ > 30\%$) were obtained (Hariharan *et al.*, 2015).

Currently, in France, there is an ongoing effort to overcome the formation of the impervious carbonate layer using mechanical methods. Bodénan *et al.* (2014) and Julcour *et al.* (2015) propose an "attrition-leaching hybrid process" that couples the typical reactive carbonation with mechanical exfoliation (for physical abrasion and removal of the carbonate passivating layer). This approach allows for nearly 80% of carbonation of mining residues, without heat activation, at 180°C and 20 bar CO_2 pressure, with a particle size fraction $< 100\,\mu$m, in less than 24 h. Addition of the standard 1 M

NaCl + 0.64 M NaHCO₃ inorganic solution, proposed by Gerdemann *et al.* (2007), improved the carbonation reaction to 70% conversion in 4 h and nearly 90% in less than 24 h. The mechanisms for carbonation of mining waste are discussed in detail in this work. It also highlights the extent of the carbonation reaction which is greatly dependent on the serpentinisation degree of the feedstock. Pasquier *et al.* (2014a, 2014b) focused on the carbonation of thermally pre-treated serpentinite at low temperature and pressure under simulated flue gas conditions. Instead of increasing global pressure and temperature, the same solid material is successively subjected to batches of gas and refreshed liquid phase. The author reports the interesting result of 64 wt.% Mg leaching and 62.5 wt.% CO_2 removal from the gas phase after 18 h under mild conditions: ~22°C and 10.5 bar.

Highfield *et al.* (2012) assessed the low energy mechano-chemical processing of serpentinite with ammonium salts, for both magnesium extraction and direct/indirect CO_2 sequestration. Dry milling of serpentinite with ammonium bisulphate (ABS) yielded 70% Mg as boussingaulite, after 3 h, according to the reaction:

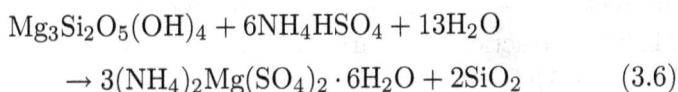

$$Mg_3Si_2O_5(OH)_4 + 6NH_4HSO_4 + 13H_2O$$
$$\rightarrow 3(NH_4)_2Mg(SO_4)_2 \cdot 6H_2O + 2SiO_2 \qquad (3.6)$$

The authors found that superficial extraction of Mg is easy, as a simple manual grinding, for 5–10 min yields up to 40% of magnesium recovery. Direct CO_2 sequestration through dry milling of $MgSO_4$ with ammonium bicarbonate (ABC) was shown to be non-viable, while a modest conversion (~35%) of $MgSO_4 \cdot 7H_2O$ was attained after 12 h milling. The same procedure, yielded 70% and 85% carbonation of MgO and Mg(OH)₂, respectively. In contrast, direct carbonation of serpentinite with ABC, besides giving low conversions, produces carbonates soluble in an aqueous mineral environment, eliminating its suitability for long-term CO_2 storage.

At last, iron oxides, that are commonly present in magnesium silicate ores, are also known to form a passive layer of hematite on the surface of the particles, as well as hindering carbonation mechanisms.

Following Huijgen and Comans (2003) suggestion, Veetil *et al.* (2015) proposed the magnetic separation of the iron oxides — prior the carbonation step — and reports 71% efficiency in the removal of iron impurities from magnesium silicate ores (serpentinites). Results for carbonation of this iron-free material have not yet been reported.

Nowadays, the direct aqueous carbonation methods still present the main challenges pointed out by the IPCC (2005) report: It requires heat treatment, fine grinding, careful solution chemistry control for long reaction times (usually many hours!) and/or extreme operating conditions (high temperature and pressure), leading to high energy penalties and unfeasible industrial processes.

3.4.4 *Indirect Aqueous Carbonation*

Although the NETL (former ARC) direct aqueous carbonation was considered the most successful route for MC, the high pressures needed (150–200 atm) and the fine grinding $<20\,\mu$m of the feedstock still stand as major energy inputs. Hence, the interest in pursuing indirect routes that allow for an independent optimisation of the Mg/Ca extraction and carbonation in two or more steps. Nonetheless, the ARC route is still an important reference and benchmark for later work.

As carbonation proceeds much faster for magnesium hydroxides than for silicates, in most of the indirect routes, the first step focuses on the conversion of magnesium silicates to hydroxides. The use of HCl for magnesium extraction from the minerals matrix was early investigated by Lackner and co-workers at the LANL (Lackner *et al.*, 1995; Butt *et al.*, 1998). However, it was soon realised that the energy required for the recovery of HCl produces more CO_2 than the process could possibly sequester (Huijgen and Comans, 2005). In order to reduce that energy penalty, Wendt *et al.* (1998) proposed a molten salt process that used $MgCl_2$ as the extraction agent. However, the process was shown to be unaffordable due to the corrosive nature of the reactants (Huijgen and Comans, 2003) and the unrealistic commercial supply of consumables (Newall *et al.*, 1999). Kakizawa *et al.* (2001), proposed the carbonation of wollastonite in the presence of acetic acid, a non-corrosive acid. Yet again, the problem stands in the additives recovery/recycling stages. More recent R&D work on enhanced extraction with acetic acid focuses on the carbonation of industrial residues such as steelmaking slags (Teir, 2008; Bobicki *et al.*, 2012). Similar to the pH swing work by Teir *et al.* (2007a, 2007b), work in South Africa by Meyer *et al.* (2014) attempted carbonation of platinum group metal (PGM) mine tailings. As a result of a large content of orthopyroxene minerals, the conversion levels were disappointing, with 30%, 3% and 9% carbonation of the material's Ca, Mg and Fe, respectively, using 2 M HCl at 70°C for extraction and addition of 15 M NaOH to accommodate carbonation. The carbonation step was 96%, 59% and 98% efficient for Ca, Mg and Fe, respectively.

At the Ohio State University, Park *et al.* (2003) addressed the optimisation of a pH swing process where the magnesium is extracted from the mineral lattice at low pH and later converted to $MgCO_3$ in an alkaline medium. In this and later work by others, complexing agents (EDTA, oxalic acid, orthophosphoric acid) and various other acids as e.g. H_2SO_4, HNO_3, KOH, NH_3, NH_4Cl, $(NH_4)_2SO_4$, NH_4NO_3, HCOOH and DL lactic acid were tested as leaching enhancing additives (Park *et al.*, 2003; Maroto-Valer *et al.*, 2005; Teir *et al.*, 2007a; Alexander *et al.*, 2007; Ghoorah *et al.*, 2014). Strong acids showed to be more effective at extracting magnesium than bases and complexing agents, making them good candidates for the pH swing process. Even so, the acid losses and the amounts of NaOH required for formation of $MgCO_3$, besides arising environmental concerns, compel a process too expensive for industrial implementation (Sipilä *et al.*, 2008).

Over the years, several authors (Botha and Strydom, 2001; Zhao *et al.*, 2010) focused on R&D of mechanisms and methods for the carbonation of $Mg(OH)_2$ slurries derived from silicate minerals and industrial wastes. Fricker and Park (2013) demonstrated the rapid and spontaneous nature of the $Mg(OH)_2$ aqueous carbonation reaction at temperatures and pressures up to 400°C, 15.2 bar. It is confirmed that MgO as a reaction intermediate should be avoided while hydrated carbonate intermediates may result in significant conversion under moderate conditions. As a continuation of this, Swanson *et al.* (2014) studied the formation of different hydrated magnesium carbonate phases while aiming at anhydrous magnesite. Following the same line of thought, Li *et al.* (2014) suggests the replacement of typical chemical sorbents (MEA amines, e.g.) by $Mg(OH)_2$ readily available from magnesium-enhanced lime flue gas desulphurisation processes. The authors present a detailed study on different operating parameters of a bubble column reactor and carbonation reaction kinetics concluding that CO_2 removal efficiency is strongly influenced by the interaction between $Mg(OH)_2$ dissolution and $MgCO_3$ precipitation.

In recent years, no noteworthy breakthroughs on the recovery of additives were accomplished and two-step aqueous carbonation routes evolved to multi-step configurations. At the University of Nottingham, UK, Wang and Maroto-Valer (2011b) proposed a pH swing process based on ammonium salts. The process is presented in Figure 3.3, and it involves five steps: (1) CO_2 from flue gases is scrubbed with chilled ammonia to form ammonium (bi)-carbonate (NH_4HCO_3: AC, or $(NH_4)_2CO_3$: AC/ABC); (2) extraction of magnesium from the silicate rock using an aqueous solution of ABS at up to 100°C to produce mainly $MgSO_4$; (3) ammonia is used to increase the pH and remove impurities (Fe, Al, Cr, Cu, etc.) in the form of hydroxides; (4) the remaining aqueous solution of $MgSO_4$ is treated with NH_4HCO_3 to precipitate hydromagnesite and finally (5) the ABS is recovered (presumably) by evaporation and

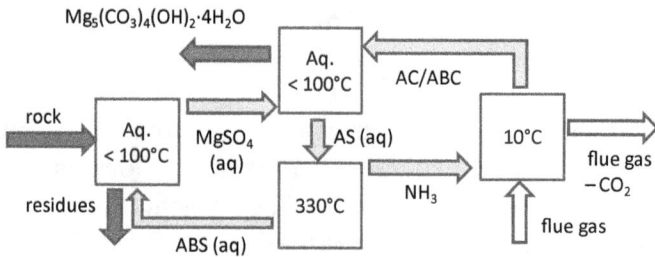

Figure 3.3 pH swing process based on ammonium salts developed at The University of Nottingham, UK (taken from Geerlings and Zevenhoven, 2013).

thermal decomposition of $(NH_4)_2SO_4$ at 330°C (Wang *et al.*, 2013). The extraction experiments showed that NH_4HSO_4 extracts 100% magnesium from serpentine at 100°C after 3 h followed by 96% of magnesium carbonation. This multi-stage pH swing method was also tested for antigorite (Sanna *et al.*, 2013) and olivine (Sanna *et al.*, 2014), as well with promising results. The total cost to sequestering 1 tonne CO_2, using serpentinite, was estimated to be $100 when using a solid/liquid ratio of 50 g/L in the extraction step. Regrettably, the recovery of reactants was not fully addressed as the recovery of ABS through thermal decomposition of AS is complicated due to its corrosive nature and shown to be energetically and economically unattractive (Wilhamson and Puschaver, 1997; Romão *et al.*, 2014). Also the process requires the handling of large streams: 4.9 tonnes of serpentine, 0.6 tonne of NH_4HSO_4, 4.7 tonnes of $(NH_4)_2CO_3$ and 16 tonnes of water are required to sequester 1 tonne of CO_2 (Wang and Maroto-Valer, 2013).

In order to minimise the energetic penalty linked to ABS regeneration, Sanna and Maroto-Valer (2014), of Heriot-Watt University, UK, suggest the replacement of ABS by a sodium bisulphate solution for extraction of magnesium. NaOH is used to capture CO_2 into Na_2CO_3 which subsequently reacts with $MgSO_4$ solution to produce $MgCO_3$ and Na_2SO_4. After carbonation, the recovery of the sodium sulphate benefits from its low solubility at low temperatures (7 g/100 mL) circumventing an evaporation step. The reported efficiency for this approach, 50% magnesium extraction at 70°C and 90% carbonation falls to the same range of the earlier results reported by Wang and Maroto-Valer (2011a, 2011b). The regeneration of Na_2SO_4 to NaOH and $NaHSO_4$ appears to be somewhat complex: It is mentioned to be achievable in three stages employing CO_2, an inorganic waste as source of CaO and an acidic environment but no further details are specified.

Balucan and Steel (2015), see also Steel and Balucan (2015), studied the use of ternary amines, of which triethyl amine (Et_3N) was found to be the best suitable, in a pressure-swing process aiming at the conversion of dissolved $MgSO_4$ (for example a tailing from a nickel mine) with aqueous dissolved CO_2 to sulphuric acid H_2SO_4

and $MgCO_3$. Process conditions are limited to temperature $<110°C$, atmospheric pressure and time <2 h. This would be applied to the high pressure acid leaching (HPAL) process for extracting nickel from laterite ores. The Et_3N was found to be able to raise pH from 2 (after Mg dissolution from serpentinites) to 10 (for $MgCO_3$ precipitation). Yet to be assessed are the energy input requirements of the concept.

By and large, the mechanisms for aqueous CO_2 mineral sequestration are fairly well understood and it is established that the dissolution rate is speed up by: Silicate ore pre-treatments/activation, application of additives and mechanic/chemical removal of the SiO_2 layer. Still, no aqueous process has yet shown to be attractive enough for developing towards pilot and/or industrial scale (Giannoulakis *et al.*, 2014).

3.4.5 *Mixed Wet/Dry Routes: Indirect Multi-stage Solid/Gas (Dry/Wet/Dry) Carbonation*

As earlier discussed, the direct carbonation of silicate minerals is not viable due to low conversions and slow reaction rates. On the other hand, the carbonation of MgO and especially $Mg(OH)_2$ was shown to occur much faster (Butt *et al.*, 1996, 1998; Béarat *et al.*, 2006). Magnesium is an element abundant in nature but mainly combined with minerals in silicate rocks. MgO and $Mg(OH)_2$ forms are rare in nature, therefore the necessity of developing a method for their production. Ergo, Zevenhoven *et al.* (2006), recommended a stepwise approach of the gas–solid MC: First the MgO is produced from serpentinite in an atmospheric reactor which is then followed by its hydration and carbonation at high pressure according to:

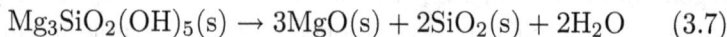

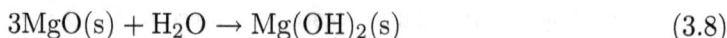

$$Mg_3SiO_2(OH)_5(s) \rightarrow 3MgO(s) + 2SiO_2(s) + 2H_2O \quad (3.7)$$

$$3MgO(s) + H_2O \rightarrow Mg(OH)_2(s) \quad (3.8)$$

$$Mg(OH)_2(s) + CO_2 \rightarrow MgCO_3(s) + H_2O \quad (3.9)$$

The initial $Mg(OH)_2$ carbonation studies showed that, at first, it proceeds quite fast but rapidly becomes limited by the build-up of a carbonate layer on the surface of the reacting particles, thus prohibiting full carbonation of $Mg(OH)_2$ (Butt *et al.*, 1996; Béarat *et al.*, 2006; Zevenhoven *et al.*, 2008). These findings were the leadoff for the construction of a test facility for carbonation studies at ÅAU.[1] Fagerlund *et al.* (2010), designed and built a lab-scale pressurised fluidised bed (PFB) reactor and initiated tests on the carbonation of MgO and Mg(OH)₂. At the same time, Nduagu (2008) worked on a method for production of Mg(OH)₂ using ammonium sulphate (AS) to extract magnesium from serpentinites, at a rock: AS mass ratio of 2:3, at 400–440°C, 30–60 min (optimised according to the rock's iron content and particle size) following:

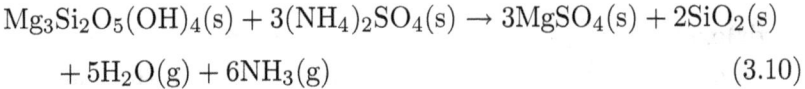

$$Mg_3Si_2O_5(OH)_4(s) + 3(NH_4)_2SO_4(s) \rightarrow 3MgSO_4(s) + 2SiO_2(s)$$

$$+ 5H_2O(g) + 6NH_3(g) \tag{3.10}$$

while more detailed studies revealed that the actual conversion reaction is most likely

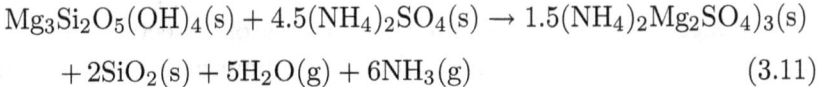

$$Mg_3Si_2O_5(OH)_4(s) + 4.5(NH_4)_2SO_4(s) \rightarrow 1.5(NH_4)_2Mg_2SO_4)_3(s)$$

$$+ 2SiO_2(s) + 5H_2O(g) + 6NH_3(g) \tag{3.11}$$

and, in water, the solid product decomposes into (dissolved) AS and MgSO₄ (Nduagu *et al.*, 2014; Highfield *et al.*, 2015).

The merging of the Mg(OH)₂ production and its subsequent carbonation (separately optimised) processes results in the multi-staged gas/solid (or three-stage dry/wet/dry) carbonation route developed at ÅA (frequently addressed in the literature as the "ÅA route") shown in Figure 3.4.

The three-stage process involves first extraction of magnesium ion, Mg²⁺, from the rock using a solid/solid reaction with AS salt, producing water-soluble magnesium sulphate. This aqueous extract is converted into magnesium hydroxide precipitate using ammonia vapour released from the solid/solid reaction. Finally the isolated magnesium hydroxide is carbonated in a PFB reactor. Besides magnesium carbonate, MgCO₃, significant amounts of iron

---

[1] In 2005, R. Zevenhoven moved from Helsinki University of Technology to ÅA University.

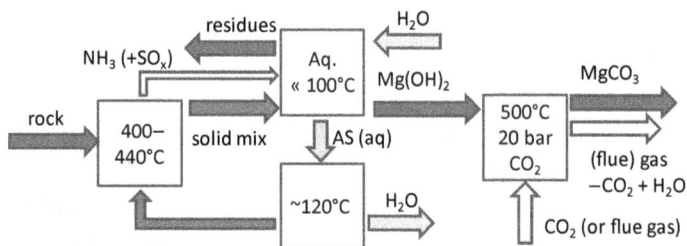

Figure 3.4 Multi-staged dry/wet/dry carbonation route developed at ÅA (taken from Geerlings and Zevenhoven, 2013).

(hydr)oxides are also produced. The heat generated by the carbonation is ~1/3 of the heat that is needed for the magnesium extraction, making it necessary to use (waste) heat from, preferably, the process that generates the CO_2. In addition, compressing the entire flue gas requires a significant power input that in large measure can be recovered by expanding the "CO_2-free" vent gas. Thus, since 2012, most work at ÅA focuses on direct carbonation with CO_2-containing gas, while for the carbonation process it has been shown experimentally that a simulated flue gas and pure CO_2 give similar carbonation results for a given CO_2 partial pressure (Slotte and Zevenhoven, 2013; Slotte *et al.*, 2013; Zevenhoven *et al.*, 2012, 2013).

A key benefit of this path is that the heat generated during the (exothermic) carbonation stage can be fed back into the (endothermic) Mg extraction stage. Experimental findings so far show that "in-house" serpentinite-derived $Mg(OH)_2$ samples perform better during carbonation, mainly as a result of a much higher specific surface area ($>40\,m^2/g$) as compared to commercial $Mg(OH)_2$ samples ($<10\,m^2/g$) (Fagerlund *et al.*, 2012a, 2012b; Stasiulaitiene *et al.*, 2011). The origin (quality) of the serpentinite determines (via its iron content) the optimal conditions for Mg extraction, but the properties of the $Mg(OH)_2$ produced are otherwise practically identical. Iron-based impurities (visible as yellow discolouring) result in a lower specific surface area and hence a lower carbonation reactivity.

$Mg(OH)_2$ produced from a Lithuanian serpentinite was carbonated up to 65%, in 15 min and 51 bar pure CO_2 (Stasiulaitiene *et al.*,

2014), while 70.3% conversion in 30 min was obtained with $Mg(OH)_2$ produced from Portuguese rock at 510°C, 20 bar pure CO_2, size fraction <74 μm (Romão, 2015; Romão *et al.*, 2016). Typical reaction temperatures for the serpentinite/AS salt reaction are somewhat lower, at 400–440°C, as to avoid decomposition and sublimation "losses" of the AS salt, as recently reported (Nduagu *et al.*, 2014). A drawback of using a fluidised bed is that it puts restrictions on the particles size: The serpentinite-derived samples typically have large amounts of particle size fractions <20 μm while most PFB tests were done with size fraction 74–125 or 125–212 μm.

With process times for $Mg(OH)_2$ production from serpentinite of the order of 30 min, it can be concluded that total process times from rock to carbonate can be accomplished in less than 1 h. However, chemical conversion levels obtained are limited. The best recovery of Mg (as carbonate) from serpentinite, expressed as "Mg extracted (as $Mg(OH)_2$) × $Mg(OH)_2$ carbonated" so far is 80% × 70%, giving 56% overall. The goal is to reach 90%×90% = 81% overall conversion levels. Note, however, that for the wet process, e.g. via hydromagnesite $4MgCO_3 \cdot Mg(OH)_2$, the maximum possible carbonation conversion of Mg is only 80%.

As regards, the salt used for the Mg extraction, recent work by Romão *et al.* (2014) has shown that ABS gives a 40% lower energy input requirement for Mg extraction than AS as used in the work reported here. However, this benefit is more than offset by an energy penalty that results from thermally converting AS into ABS for re-use. For the recovery of the AS salt from aqueous solutions, the use of mechanical vapour recompression (MVR) was suggested and analysed by Björklöf (2010) to be an energy efficient alternative for boiling off the water; more recently the use of reverse osmosis was tested for this purpose (Virtanen, 2015).

In a recent case study for application of this route (without CO_2 pre-separation) to an industrial lime kiln, it was found that the heat input requirements were completely satisfied by waste heat from the lime kiln, while a power input of 0.9 MJ/kg CO_2 fixed was determined primarily by the flue gas (kiln product gas) compression stage (Slotte *et al.*, 2013; Slotte and Zevenhoven, 2013). The same case study was earlier reported for an application to CO_2 pre-separated from flue gas (Romão *et al.*, 2012a).

3.4.6 *Mixed Wet/Dry Routes: Indirect Multi-stage Solid/Gas (Dry/Wet/Wet) Carbonation*

One disadvantage of the conventional ÅA route is the somewhat complicated sequence involving a hot solid/solid reactor (Mg^{2+} extraction), followed by aqueous precipitation reactors (Mg and Fe hydr(oxide) production), and finally a PFB gas/solid reactor (carbonation). For this reason, an "alternative" ÅA route is being explored that can also operate directly on flue gas (see Figure 3.5). Here, the precipitation and carbonation steps are coupled in a low temperature aqueous solution process in which the primary reaction intermediate (a solution of Mg^{2+} and SO_4^{2-} ions) is carbonated directly by addition of NH_4OH (recycled NH_3 as before) while sparging with flue gas. Depending on temperature, the products are magnesium (hydrocarbonates), primarily nesquehonite ($MgCO_3 \cdot 3H_2O$) and hydromagnesite ($4MgCO_3 \cdot Mg(OH)_2$), besides iron(hydr)oxides.

The chemistry for this process scheme involves Reactions (3.10), (3.11), as mentioned previously, followed by, at near ambient conditions, in aqueous solution:

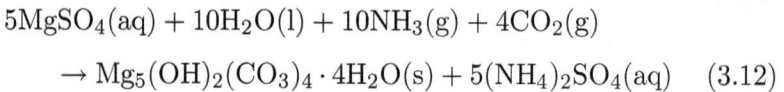

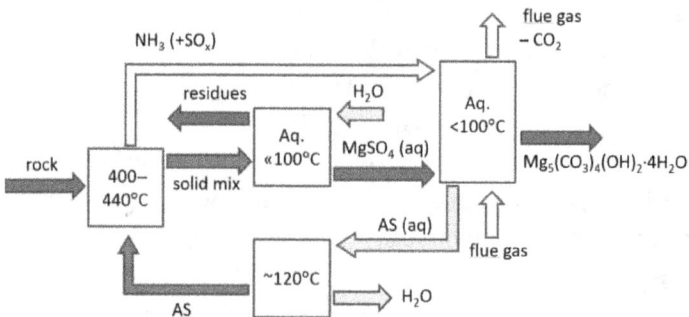

$$5MgSO_4(aq) + 10H_2O(l) + 10NH_3(g) + 4CO_2(g)$$
$$\rightarrow Mg_5(OH)_2(CO_3)_4 \cdot 4H_2O(s) + 5(NH_4)_2SO_4(aq) \quad (3.12)$$

Figure 3.5 Multi-staged dry/wet/wet carbonation route developed at ÅA (taken from Zevenhoven *et al.*, 2015).

Note that the "alternative" ÅA route involves a simpler CO_2 contacting stage (to a low temperature aqueous solution) compared to the route developed in the UK, by Maroto-Valer and co-workers, discussed above and shown in Figure 3.3.

This route combines the precipitation and carbonation of magnesium ion into one step, thus simplifying the carbonation process. When all unreacted rock, SiO_2 and Fe-(hydr)oxides have been removed from the stream, the $MgSO_4$ solution is exposed to a constant stream of CO_2-containing flue gas. The reaction between $MgSO_4$ and CO_2 cannot proceed at the acidic pH typical of the extract. However, addition of NH_3 raises the pH sufficiently to initiate precipitation of magnesium (hydro)carbonates. The AS produced is recovered and recycled back to the extraction step (as in the "conventional" ÅA route). Nesquehonite ($MgCO_3 \cdot 3H_2O$) and lansfordite ($MgCO_3 \cdot 5H_2O$) are typically formed at temperatures (CO_2 pressures) in the range 30°C (0.01 bar CO_2) to 50°C (1 bar CO_2). Above 50°C hydromagnesite ($Mg_5(CO_3)_4(OH)_2 \cdot 4H_2O$), starts to form (Hill *et al.*, 1982).

Tests were made starting with synthetic $MgSO_4$ solutions (Åbacka, 2013) followed by $MgSO_4$ solutions produced from (Finnish) serpentinite rock. The full experimental set-up with a lab-scale rotary kiln (Carbolite HTR11/75), and typical solution pH values, is shown in Figure 3.6.

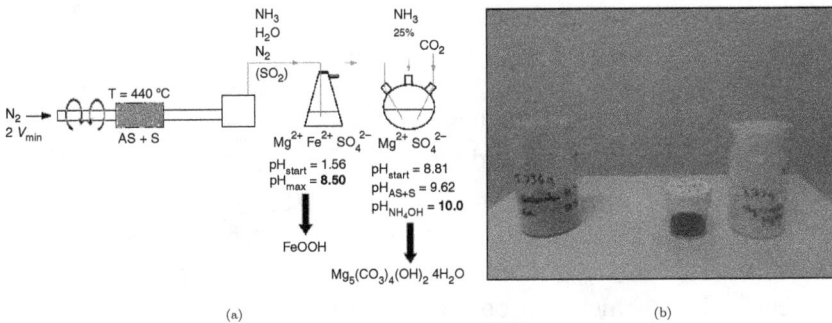

Figure 3.6 (a) Experimental set-up for separate FeOOH precipitation and Mg carbonation using ammonia from AS/S reaction; (b) Solid products from Fe and Mg precipitated/carbonated together (one sample) versus separately precipitated FeOOH and carbonated Mg (two samples).

Experiments were done with or without separation of iron before sparging the solution with CO_2. The products obtained were either iron carbonate (siderite) mixed with hydromagnesite, or iron oxy-hydoxide (goethite) separate from hydromagnesite, respectively — see the product samples in Figure 3.6(b). pH = 10 is suitable for hydromagnesite precipitation; at the end of the experiments the pH had decreased to 8.0–8.2. The combined (Fe present) magnesium carbonation efficiency varied from 53 to 88%, while for the carbonation of the iron-free solution, the magnesium carbonation efficiency was 72. More detail on the comparison between the performances and energy efficiencies of the two ÅA routes is given elsewhere (Zevenhoven *et al.*, 2015), besides what is given in the next section.

3.5 Energy Efficiency of Magnesium Silicate Carbonation Processing

It isn't surprising that "energy use" is a crucial factor for sustainable and economically feasible CCS, as it is a response to what is primarily an energy problem: The global infrastructure for producing electricity, fuel and heat is strongly dependent on fossil carbon-derived fuel. Therefore, any use of electricity or heat in a CCS process can be directly re-calculated to CO_2 emissions arising from that energy input, leading to the concept of "avoided CO_2 emissions". For CCGS, this is directly related to the energy needed for regenerating the CO_2 capture medium and the compression of CO_2 to storage pressure and also overcome transport line pressure drop. As reported earlier (Rubin *et al.*, 2007; Zevenhoven *et al.*, 2008; Zevenhoven and Fagerlund, 2010), typical values for CO_2 produced when generating heat or electricity are 0.2 and 0.6 kg/kWh, respectively. With these values, a maximum allowable value for energy use for CO_2 sequestration can be given: 18 MJ heat/kg CO_2, or 6 MJ (= 1.67 kWh) electricity/kg CO_2. Exceeding this results in a "zero game" with respect to CO_2 which can be motivated only by high market values for the solid products.

In most work on CO_2 mineralisation process concepts reported in the open literature, energy is not mentioned. Although the energy related to CO_2 pre-separation could be taken out of the equation (even though most MC studies consider a pure CO_2 feed), energy input requirements for MC can involve:

- Mining, crushing and grinding of material: Here an overburden or tailing from mining for another purpose is beneficial, besides a low Bond index that quantifies the grinding energy needs, a high purity with respect to (here) magnesium as well as the preferable crystal form, while finally the particle size must be optimised against reactivity.
- Transport: Typically transport costs are low for fossil fuels and many other mined materials. Nevertheless, long distances of more than a few 100 km for rock transport to a CO_2 source should be avoided, especially on land.
- Net heat and power for the actual carbonation process: These should be compensated as much as possible by waste heat from the CO_2 generating process, while the use of power (i.e. electricity) should be minimised. Nonetheless, in some cases electrochemical steps are part of MC processes. "Parasitic" CO_2 generation for a certain amount of heat is typically only 1/3 (depending of on temperature) of that for the same energy as electricity: See Appendix in Zevenhoven and Fagerlund (2010) or Björkström and Zevenhoven (2012).
- Post-processing and other energetic effects: Upgrading of product quality, drying or products, recovery of chemical additives, besides transport of residue back to rock source mine.

These will be addressed in following sections, followed by a brief analysis on water use and life cycle assessment (LCA) studies related to CO_2 mineralisation.

3.5.1 *Mining, Crushing, Grinding*

This was briefly addressed above (Section 3.4.1), mentioning 50–250 μm as typical suitable particle sizes used by most researchers. Besides the external surface, the internal surface (i.e. pore walls) of particles to be carbonated are important for solid/fluid conversion rates: Fagerlund *et al.* (2012a) reported much higher degrees of conversion for serpentinite-derived $Mg(OH)_2$ with specific surfaces of 40–50 m^2/g compared to a (commercial) sample with <10 m^2/g. A corresponding (non-porous) particle would, assuming a density of 2,500 kg/m^3 and a spherical shape, have to have a size <0.1 μm. The ARC work (Gerdemann *et al.*, 2007; O'Connor *et al.*, 2005) gives ∼15 kWh (48 MJ)/t for crushing and grinding to <75 μm, increasing to ∼85 kWh (300 MJ)/t for further grinding to <38 μm. Romão *et al.* (2012b) mentions 150 kJ/kg CO_2 fixed which, with ∼3 kg rock needed per kg CO_2, corresponds to 50 MJ/t rock. For (stone) mining, the energy use was recently reported to lay in the similar range 49–371 MJ/t, with a typical average of 188 MJ/t (Kirchofer *et al.*, 2012). The crushing and grinding of ultramafic rock, to be spread out in the environment at a location where (enhanced) natural weathering will then commence was found by Moosdorf *et al.* (2014) to be a dominating factor for the CO_2 sequestration efficiency. Crushing rock down to sizes of ∼1 mm was estimated to require 5–10 MJ/t rock. Finer grinding to realise complete weathering within 1 year would require more grinding energy, leading to CO_2 emissions in the range 0.07–0.22 tCO_2/t rock processed. For mining an energy requirement of 18.8 MJ/t rock is given, significantly less than what others report.

Thus, with similar values for mining and for crushing and grinding a combined 90–670 MJ ≈ 25–185 kWh/t rock would be needed: A typical power plant would generate 15–110 kg CO_2 while producing this electricity. This corresponds to 4–33% of the CO_2 that could be sequestered by 1 tonne of rock.

Particle size has an effect on the rate and eventual final extent of the carbonation reactions. Yeo *et al.* (2015) concluded that below a certain particle size, thermal or chemical treatments are more effective than a further reduction in particle size. Grinding a magnesium silicate from 75 μm (d$_{80}$) to 37 and 10 μm increases the extent of reaction from 16 to 61% and finally 81%, while grinding energy increases from 11 kWh/t rock to 81 and 231 kWh/t rock. Corresponding CO_2 emissions for processing a tonne of rock would increase from 4 to 93 kg CO_2 (assuming electricity from natural gas). Moreover, the products (a silica and carbonate mix) may become too fine for use as a concrete aggregate. Instead of grinding from 75 (d_{80}) to 19 μm, the same extent of reaction (81%) can be obtained by thermal treatment of 1.2 GJ/t rock. Using natural gas to supply this heat would produce 60 kg CO_2. Yet better results are obtained (95% conversion) by combining grinding to 75 μm with acid leaching. The energy needs (heat) for chemicals regeneration may be as high as 5 GJ/t rock processed, depending on the extent of acid regeneration.

3.5.2 *Transport*

Transportation of large amounts of material across the planet is one of the largest sectors of the global economy. Raw materials, fossil fuels and industrial products and half-products cover long distance by water, road, train and air. One feature of bulk transport of non-fragile and non-hazardous materials is the rather low cost, largely the result of relatively small energy needs. Kirchofer *et al.* (2012) give requirements of the order of 250 kJ/t material per km for transport by truck until 100 km and by train above that (for very large distances ship transport may be more attractive). This value gives for a transport distance of, say, 200 km, an energy input requirement of 50 MJ per tonne rock, which is of the same order as the lower limit for mining or crushing and grinding. Generating about 50 MJ to 14 kWh electricity would give 8.4 kg CO_2, which is ~2% of the CO_2 that can be sequestered by one of the rocks suitable for MC. Likewise, Moosdorf *et al.* (2014) estimate that CO_2 emissions from rock transport are 0.5–3% of the CO_2 that is eventually sequestered, even for transport distances of several 1,000 km.

3.5.3 *Net Heat and Power for the Actual Carbonation Process*

Typical for material processing where a chemical reaction is one of the process steps, is that the energy effect of this determines the overall energy use and energy efficiency picture. (Exceptions are typically processes operating near ambient conditions.) Heat and power (electricity) is needed for pre-heating, pressurising and transporting ingoing process streams, fulfilling heat requirements of the reactor (or reactors) and guarantee mixing of species, and for creating the driving forces for downstream separations and solid product upgrading. Equally important is process heat integration based on heat recovery from product stream and optimised heat exchange.

As mentioned, the energy input/output characteristics of MC processes are seldom addressed in scientific reporting. One theoretical advantage of MC is that the overall process chemistry is heat-generating, while in practice this is difficult to make use of since MC processes are usually a combination of heat-generating and heat-requiring steps taking place at different temperatures. For a proper assessment of this, the quality of heat, i.e. its temperature T, must be considered with respect to surroundings at temperature $T°$. Thus, the so-called exergy of heat Q (conceptually equal to the maximum amount of work, or electricity that can be produced from it) is defined as $Ex(Q) = Q \times (1 - T/T°)$, while the exergy of power P equals $Ex(P) = P$. (This can be readily extended to include pressure with respect to the surroundings pressure, $p°$, or even further, to include the chemical exergy through comparing each element with its state in the surrounding environment and mixing energy (Szargut *et al.*, 1988; Björklöf and Zevenhoven, 2012)).

In addition to the goal of accomplishing MC with rock material amounts that approach the minimum needs, the second goal is to accomplish MC with energy input requirements as close as possible to thermodynamic limits for the conditions where chemical kinetics are considered acceptable. For the ÅA route discussed above, operating as a stand-alone process supplied with pure CO_2, assuming 95% extraction of Mg from serpentinite followed by 95% carbonation to $MgCO_3$ gives an overall exergy requirement of 4.33 MJ/kg CO_2 (Romão *et al.*, 2012b), in the form of heat of 400–500°C. This can be compared with a later analysis of an integrated process at a lime kiln, operating on compressed flue gas, where heat and power input requirements were found to be 2.6 MJ/kg CO_2 and 0.9 MJ/kg CO_2 respectively, where the heat requirements can be covered by waste heat available from the lime kiln process (Slotte *et al.*, 2013). In a more recent analysis, both the ÅA route and the "alternative" ÅA route (see Figure 3.6) were applied to the lime kiln (in Finland) and a natural gas-fired combined cycle (NGCC) plant (in Singapore). The flue gas feed pressures for the (conventional) ÅA route were 80 and 400 bar, respectively, and dry CO_2 concentrations 23.3 and 5.2 vol%. The results are summarised in Table 3.2 (Zevenhoven *et al.*, 2015).

The findings tabulated in Table 3.2 show that the dry carbonation route with lime kiln gas has the lowest exergy demand of the processes considered, and significantly better than the wet-process route. However, the wet route has a zero exergy input as power, requiring only (waste) heat. Thus, for both lime kiln cases, the heat required can be obtained from the kiln gas while the wet process route does not require flue gas compression.

The earlier work at ARC (currently NETL; Gerdemann *et al.*, 2007; O'Connor *et al.*, 2005) includes a thermal treatment (at 600–725°C) of ~300 kWh/t rock for the two serpentinites considered. Using a temperature of 725°C, 300 kWh (=1.08 GJ) correspond to an exergy of 0.77 GJ/t rock. With 3–4 tonnes rock needed per tonne CO_2, this implies 2.3–3.1 GJ/tCO₂ sequestered.

In Geerlings and Zevenhoven (2013), a brief energy efficiency assessment was also made of the "Nottingham route" given in Figure 4.4. In that process, regeneration of the ABS for re-use requires a heat-driven conversion of the dissolved AS to NH_3 (for CO_2 capture re-use) and ABS at 330°C. This latter step puts an energy penalty on the process on the order of 1.3 kWh = 4.7 MJ/kg CO_2 at 330°C, i.e. an exergy requirement of 2.4 MJ/kg CO_2, at surroundings T° = 15°C. (Note that the AS salt that is be regenerated to ABS is dissolved in water.) The extraction, capture, and carbonation stages produce 1.5 kWh = 5.4 MJ/kg CO_2 at 80–90°C which implies an exergy generation of 1.0–1.1 MJ/kg CO_2 (Sanna *et al.*, 2013). A heat pump may be used to raise the temperature of the generated heat to the level needed for the ABS salt regeneration.

For MC processes that use electrochemical steps, this generally implies the use of electrolysis (and fuel cells) for producing aqueous hydrochloric acid and sodium hydroxide reactant solutions. These then allow for the necessary pH swing for CO_2 mineralisation. Results from three recent publications (House *et al.*, 2007; Li *et al.*, 2009;

Table 3.2 Process simulation results for two ÅA carbonation routes and two flue gas types (taken from Zevenhoven et al., 2015).

	Serpentinite Rock in, kg/h (T°C)	Flue Gas in, kg/h (T°C)	Mg-Product Out*, kg/h (T°C)	Flue Gas Out, kg/h (T°C)	CO_2 Bound, (T°C)	Exergy Input Heat kW	Exergy Input Power kW	Exergy Input Heat, MJ/kg CO_2 Bound	Exergy Output Power, MJ/kg CO_2 Bound
Dry carbonation, lime kiln	550 (15)	620 (500)	386 (150)	489 (92)	187	135§	46	2.6(§)	0.89
Dry carbonation, natural gas PP	550 (15)	2,427 (500)	385.8 (150)	2240.6 (18)	187	691#	439	11.44#	8.27
Wet carbonation, lime kiln	550 (15)	620 (500)	353 (30)	400 (30)	190	814$	0	15.4$	0
Wet carbonation, natural gas PP	550 (15)	2,426 (500)	353.2 (30)	2294.2 (30)	189	635#	0	12.1#	0

*$MgCO_3$ for dry carbonation, $Mg_5(CO_3)_4(OH)_2 \cdot 4H_2O$ for wet carbonation.

§This heat can be obtained from 28,620 kg/h lime kiln flue gas, lowering the temperature from 500 to 380°C.

$This heat can be obtained from 39,400 kg/h lime kiln flue gas, lowering the temperature from 500 to 362°C.

#This heat can be obtained from 28,500 kg/h flue gas, lowering the temperature of the gas from 500 to 380°C (*It is assumed that while 2,427.4 kg/h flue gas is processed in the carbonation process, a larger flue gas of at least 28,500 kg/h is produced by the power plant.*).

Teir *et al.*, 2007a, 2009) on the subject that come to different conclusions, are used in an assessment by Björklöf and Zevenhoven (2012). The exergy concept was used to show that enthalpies and process heat values cannot simply be added up, regardless of temperature, in order to calculate an overall value. Compared to process steps that generate or require moderate amounts of heat, the input of electricity for the electrolysis puts a large energy penalty on the process, even if some of this can be generated with a fuel cell. The electrochemical steps must achieve efficiencies around 90% for the overall process to give a positive overall capture efficiency and process routes that use heat integration can more easily offer the process energy efficiency that is required.

In recent work from Singapore reported by Hemmati *et al.* (2014a, 2014b), bipolar membrane electrodialysis was used, and the major goal was producing valuable solid products. No CO_2 sequestration efficiency is given (as CO_2 fixed versus CO_2 produced when generating the electricity needed for the electrodialysis). An exergy analysis would quantify whether the process is carbon-negative.

3.5.4 *Post-Processing and Other Energetic Effects*

Post-processing is addressed by Kirchofer *et al.* (2012), reporting that at the MC process itself (downstream processing), the energy penalty is very small (<3 MJ/t solids). For transport, the energy requirements are similar to that of the rock used in the process (see the numbers given above) while for backfill disposal, 50% of the lower value for mining can be used, say 20–50 MJ/t rock.

3.5.5 *LCA and Water Use*

Probably the best picture on the benefit of CCS methods or for comparing different CCS/CCUS methods is obtained by LCA. Energy use of processes can be immediately re-calculated into new CO_2 emissions, quantified as the global warming potential (GWP), but also effects on human health, effects on ecosystem quality, resources depletion, acidification potential, ozone layer depletion and eutrophication potential are environmental impact categories that can be considered. A handful of recent LCA studies have addressed CO_2 mineralisation. Most of them have the ÅA route given in Figure 3.5 as one of the case studies, partly because it is one of the CCS/CCUS processes for which reporting gives sufficient detail for making an LCA study. Khoo *et al.* (2011a) report on MC in Singapore, applying the ÅA route to a NGCC power plant using serpentinite (assuming a price of US\$ 7/t) imported from Australia. Calculated cases involve 90 and 100% Mg carbonation efficiency, and amine scrubbing for CO_2 pre-capture or operation on (pressurised) flue gases directly. The best results, obtained without CO_2 pre-capture, given an avoided 66 or 55% CO_2 emissions for 100 and 90% carbonation efficiency for the MC process, which drops to 58 and 42%, respectively, when the entire life cycle including shipping and mining is taken into account. Estimated costs in the range US\$ 71–159/tCO_2 may be re-calculated to US\$ 18–40/t solid products for land reclamation. Kirchofer *et al.* (2012) gave an LCA of aqueous mineralisation applied to 11 alkaline feedstocks, being five olivine, one serpentine and five industrial by-products used for 1,000 t/day CO_2 sequestration. The lowest CO_2 emissions (267 kg/tCO_2 processed) are obtained with 10 μm olivine particles reacted at 155°C during 24 h; for 4 μm serpentine the CO_2 emissions after 24 h are 654 kg/tCO_2 processed primarily because the slower dissolution gives only 71% conversion. The paper also gives values on the energy needs of the various upstream and downstream process steps.

Nduagu *et al.* (2012c) compared the ÅA route with the ARC direct MC process developed earlier in the US (see Section 3.3.2) using exergy analysis and LCA, for a pure CO_2 feed. The energy input requirements for the processes were found to be very similar at ~3.5 GJ/tCO_2, but the LCA showed a smaller environmental impact of the ÅA. Avoided CO_2 emissions are 317 and 483 kg/t mineralised CO_2 for the ARC and ÅA routes, respectively, as a result of better recoverability of additional chemicals, lower temperatures and better heat integration for the latter. Stasiulaitiene *et al.* (2013) applied the ÅA route to serpentinite from Lithuania and produced an LCA that compares it with CCGS. It is concluded that with a CO_2 capture step, mineralisation of CO_2 cannot, from an LCA point of view, compete with underground storage of CO_2. The numbers reported for MC agree with Nduagu *et al.* (2012c). More recently, Giannoulakis *et al.* (2014) reported an LCA analysis comparing both the ÅA route and the ARC work with geological storage of CO_2, for both pulverised coal firing and NGCC. It was concluded that the "life cycle greenhouse gas reduction" achievable with mineralisation is less than with geological storage as a result of energy and chemical additives use, besides the health issues related to particulate formation during mining. However, Khoo *et al.* (2011) is not mentioned, nor is other reporting (e.g. from ÅA) that specifically mentions that operating on flue gas, without CO_2 pre-separation, is a large benefit of CO_2 mineralisation compared to CCGS.

Bodénan *et al.* (2014) include LCA in the Carmex project work (see Section 3.3.2.3), distinguishing direct aqueous MC of olivine as such, with organic ligands and with mechanical exfoliation, and compare that with a no CCS case and with CCGS. CO_2 is assumed to be transported over a distance of 300 km from a coal-fired power plant to the storage site. Of five LCI categories considered, climate change (GWP) decreases in compared to a no CCS scenario while four other LCI categories (resources depletion, non-renewable primary energy use, terrestrial acidification and photochemical oxidation) the environmental impact of *ex-situ* MC is higher. Not surprisingly, the outcome depends much on what process route is chosen.

Recently, an analysis including LCA of CCS and CCU (the later including CO_2 mineralisation) was presented by Cuéllar-Franca and Azagapic (2015). GWP is the main LCI category considered yet others are touched upon as well. The work reported by Khoo *et al.* (2011a, 2011b) and Nduagu *et al.* (2012c) as mentioned above gives three of the CCU studies considered, all producing $MgCO_3$ from (process intermediate) $Mg(OH)_2$. The assessment gives CO_2 emissions in the range 524–1073 kg CO_2/tCO_2 fixed as minerals, depending primarily on upstream CO_2 capture (or not) and CO_2 concentration, and the heat source/heat integration. Within the

CCS/CCU portfolio, MC is reported to have a larger GWP than "conventional CCS" (i.e. CCGS) and CCS + EOR but a lower GWP than CCU that involves biodiesel production from algae or production of dimethyl carbonate (DMC). For other LCI categories, the environmental impacts are generally higher for CCU than for CCS, except for some LCI for biodiesel production from algae. Altogether, for CCU much depends on the eventual use of products and especially the duration of the "storage" it materialises.

A final aspect to be addressed is the water use during MC processing. For the "ÅA route" this was studied by Romão and Grigaliunaite (2013) for serpentinite carbonation. Since such rock typically contains 10–15 wt.% crystal water, there is some allowance or losses during filtration and other processing. The high solubility of ammonium and sulphate in water, on the other hand, gives the risk of losses of these: As, being used as a "flux" for extracting magnesium from the rock should be recovered for re-use at close to 100%. Water amounts used should be kept to a minimum, although washing and drying solid products and residues may require additional water processing. Also, the iron content of the rock has great impact on water amounts necessary. Settlers may be used for removing solid particles and precipitates from the aqueous streams. Filters may be necessary for final dewatering.

The picture becomes different if hydromagnesite ($Mg_5(CO_3)_4$ $(OH)_2 \cdot 4H_2O$) or even more if nesquehonite ($MgCO_3 \cdot 3H_2O$) is the final product. Serpentin(it)e brings 2 mol water per 3 mol Mg into an MC process, which is not sufficient for the water content of hydromagnesite, let alone for nesquehonite as the final product. The water consumption of an MC process will be even higher when olivines or other rock without crystal water is the feedstock material. Overall, the issue of water use and consumption deserves a serious separate study.

3.6 Acknowledgements

The authors wish to acknowledge Dr. Hannu-Petteri Mattila (currently at Paroc, Finland), Martin Slotte and Jacob Åbacka (currently at Wärtsilä Moss, Norway) of Åbo Akademi University, Dr. James Highfield of ICES/A*Star, Singapore, and Prof. Licínio M. Gando-Ferreira of the University of Coimbra, Portugal, for contributing to the work reported in this manuscript.

References

Åbacka, J. (2013). MSc thesis. *Low Temperature Carbonation of Magnesium Hydroxide and Sulphate*. Åbo Akademi University, Åbo/Turku Finland.

Ahrens, C. D. (1985). *Meteorology Today: An Introduction to Weather, Climate, and the Environment*. West Publishing Company.

Alexander, G., M. Maroto-Valer and P. Gafarova-Aksoy (2007). Evaluation of reaction variables in the dissolution of serpentine for mineral carbonation. *Fuel* 86(1–2): 273–281.

Allen, D. J. and G. F. Brent (2010). Sequestering CO_2 by mineral carbonation: Stability against acid rain exposure. *Environmental Science & Technology* 44(7): 2735–2739.

APA (2012). Roteiro Nacional de Baixo Carbono. Análise técnica das opções de transição para uma economia de baixo carbono competitiva em 2050 APA, Amadora.

Armstrong, K. and P. Styring (2015). Assessing the potential of utilization ans storage strategies for post-combustion CO_2 emissions reduction. *Frontiers in Energy Research* 3(8): 1–9.

Azdarpour, A., M. Asadullah, E. Mohammadian, H. Hamidi, R. Junin and M. A. Karaei (2015). A review on carbon dioxide mineral carbonation through pH-swing process. *Chemical Engineering Journal* 279: 615–630.

Baláž, P., E. Turianicová, M. Fabián, R. A. Kleiv, J. Brianèin and A. Obut (2008). Structural changes in olivine (Mg, Fe)2SiO4 mechanically activated in high-energy mills. *International Journal of Mineral Processing* 88(1–2): 1–6.

Balucan, R. D. and B. Z. Dlugogorski (2013). Thermal Activation of Antigorite for Mineralization of CO_2. *Environmental Science & Technology* 47(1): 182–190.

Balucan, R. D., B. Z. Dlugogorski, E. M. Kennedy, I. V. Belova and G. E. Murch (2013). Energy cost of heat activating serpentinites for CO_2 storage by mineralisation. *International Journal of Greenhouse Gas Control* 17: 225–239.

Balucan, R. D. and K. M. Steel (2015). A regenerable precipitant-solvent system for CO_2 mitigation and metals recovery. *International Journal of Greenhouse Gas Control* 42: 379–387.

Béarat, H., M. J. McKelvy, A. V. G. Chizmeshya, D. Gormley, R. Nunez, R. W. Carpenter, K. Squires and G. H. Wolf (2006). Carbon sequestration via aqueous olivine mineral carbonation: Role of passivating layer formation. *Environmental Science & Technology* 40(15): 4802–4808.

Björklöf, T. (2010). MSc thesis. *An Energy Efficiency Study of Carbon Dioxide Mineralization*. Åbo Akademi University, Turku, Finland.

Björklöf, T. and R. Zevenhoven (2012). Energy efficiency analysis of CO_2 mineral sequestration in magnesium silicate rock using electrochemical steps. *Chemical Engineering Research and Design* 90: 1467–1472.

Blencoe, J. G., D. A. Palmer, L. M. Anovitz and J. S. Beard (2012). Carbonation of metal silicates for long-term CO_2 sequestration. U.S. Patent 2012/0128571.

Blencoe, J. G., D. A. Palmer, L. M. Anovitz and J. S. Beard (2004). Carbonation of calcium silicates for long-term CO_2 sequestration. Patent WO 200409043.

Boavida, D., J. Carneiro, R. Martinez, M. van den Broek, A. Ramirez, A. Rimi, G. Tosato and M. Gastine (2013). Planning CCS Development in the West Mediterranean. *Energy Procedia* 37: 3212–3220.

Bobicki, E. R., Q. Liu, Z. Xu and H. Zeng (2012). Carbon capture and storage using alkaline industrial wastes. *Progress in Energy and Combustion Science* 38(2): 302–320.

Bobicki, E. R., Q. Liu, Z. Xu, E. R. Bobicki, Q. Liu and Z. Xu (2014). Ligand-promoted dissolution of serpentine in ultramafic nickel ores. *Minerals Engineering* 64: 109–119.

Bobicki, E. R., Q. Liu and Z. Xu (2015). Mineral carbon storage in pre-treated ultramafic ores. *Minerals Engineering* 70: 43–54.

Bodénan, F., F. Bourgeois, C. Petiot, T. Augé, B. Bonfils, C. Julcour-Lebigue, F. Guyot, A. Boukary, J. Tremosa, A. Lassin, E. C. Gaucher and P. Chiquet (2014). *Ex situ* mineral carbonation for CO_2 mitigation: Evaluation of mining waste resources, aqueous carbonation processability and life cycle assessment (Carmex project). *Minerals Engineering* 59: 52–63.

Boerrigter, H. (2010). A process for preparing an activated mineral. EP2242723: A1.

Botha, A. and C. A. Strydom (2001). Preparation of a magnesium hydroxy carbonate from magnesium hydroxide. *Hydrometallurgy* 62(3): 175–183.

BP (2015). *BP Statistical Review of World Energy 2015.*

Brent, G. F. *et al.* (12 co-authors) (2015). Mineral carbonation of serpentinite: From the laboratory to pilot scale — the MCi project. Presented at ACEME2015. New York City, NY, USA, June 2015, p. 10.

Bundeleva, I. A., B. Ménez, T. Augé, F. Bodénan, N. Recham and F. Guyot (2014). Effect of cyanobacteria Synechococcus PCC 7942 on carbonation kinetics of olivine at 20°C. *Minerals Engineering* 59: 2–11.

Butt, D. P., K. S. Lackner and C. H. Wendt (1998). The kinetics of binding carbon dioxide in magnesium carbonate, *Proceedings of the 23th International Conference on Coal Utilization and Fuel Systems*. Clearwater, Florida, USA.

Butt, D. P., K. S. Lackner, C. H. Wendt, S. D. Conzone, H. Kung, Y. Lu and J. K. Bremser (1996). Kinetics of thermal dehydroxylation and carbonation of magnesium hydroxide. *Journal of the American Ceramic Society* 79(7): 1892–1898.

Carey, J. W., E. P. Rosen, D. Bergfeld, S. J. Chipera, D. A. Counce, M. G. Snow, H. Ziock and G. D. Guthrie (2003). Experimental studies of the serpentine carbonation reaction. *28th International Technical Conference on Coal Utilization & Fuel Systems*, Clearwater, FL, USA, 2003, pp. 331–340.

Carneiro, J. F., D. Boavida and R. Silva (2011). First assessment of sources and sinks for carbon capture and geological storage in Portugal. *International Journal of Greenhouse Gas Control* 5(3): 538–548.

Cuellar-Franca, R. and A. Azapagic (2015). Carbon capture, storage and utilisation technologies: An analysis and comparison of their life cycle environmental impacts. *Journal of CO_2 Utilization* 8: 82–102.

Dlugogorski, B. Z. and R. D. Balucan (2014). Dehydroxylation of serpentine minerals: Implications for mineral carbonation. *Renewable and Sustainable Energy Reviews* 31: 353–367.

EC (2009). Directive 2009/31/EC of the European Parliament and of the council on the geological storage of carbon dioxide and amending Council Directive 85/337/EEC, European Parliament and Council Directives 2000/60/EC, 2001/80/EC, 2004/35/EC, 2006/12/EC, 2008/1/EC and Regulation (EC) No 1013/2006.

Eikeland, E., A. B. Blichfeld, C. Tyrsted, A. Jensen and B. B. Iversen (2015). Optimized Carbonation of Magnesium Silicate Mineral for CO_2 Storage. *ACS Applied Materials & Interfaces* 7(9): 5258–5264.

Ellsworth, W. L. (2013). Injection-Induced Earthquakes. *Science* 341(6142): 1225942-1–1225942-7.

EPA (2014). Climate change indicators in the United States, 2014. Third edition. EPA 430-R-14-004. www. epa.gov/climatechange/indicators.

Fagerlund, J. (2012). Dr. Tech Thesis. *Carbonation of Mg(OH)2 in a Pressurized Fluidised Bed for CO_2 Sequestration*. Åbo Akademi University, Åbo/Turku, Finland.

Fagerlund, J., E. Nduagu, I. Romão and R. Zevenhoven (2010). A stepwise process for carbon dioxide sequestration using magnesium silicates. *Frontiers of Chemical Engineering in China* 4(2): 133–141.

Fagerlund, J., E. Nduagu, I. Romão and R. Zevenhoven (2012a). CO_2 fixation using magnesium silicate minerals part 1: Process description and performance. *Energy* 41(1): 184–191.

Fagerlund, J., J. Highfield and R. Zevenhoven (2012b). Kinetics studies on wet and dry gas-solid carbonation of MgO and Mg(OH)2 for CO_2 sequestration. *RSC Advances* 2(27): 10380–10393.

Fauth, D. J., P. M. Goldberg, J. P. Knoer, Y. Soong, W. K. O'Connor, D. C. Dahlin, D. N. Nilsen, R. P. Walters, K. S. Lackner, H. Ziock, M. J. McKelvy and Z. Chen (2000). Carbon dioxide storage as mineral carbonates, Preprints of symposia. *American Chemical Society, Division Fuel Chemistry* 708–712.

Feron, P. H. M. and C. A. Hendriks (2005). CO_2 Capture Process Principles and Costs — Les différents procédés de capture du CO_2 et leurs coûts. *Oil & Gas Science and Technology* — Rev. IFP 60(3): 451–459.

Fricker, K. J. and A. A. Park (2013). Effect of H_2O on $Mg(OH)_2$ carbonation pathways for combined CO_2 capture and storage. *Chemical Engineering Science* 100: 332–341.

Gadikota, G., J. Matter, P. Kelemen and A.-H. A. Park (2014a). Chemical and morphological changes during olivine carbonation for CO_2 storage in the presence of NaCl and NaHCO₃. *Physical Chemistry Chemical Physics* 16(10): 4679–4693.

Gadikota, G., E. J. Swanson, H. Zhao and A.-H. A. Park (2014b). Experimental design and data analysis for accurate estimation of reaction kinetics and conversion for carbon mineralization. *Industrial and Engineering Chemistry Research* 53: 6664–6676.

Gadikota, G., C. Natali, C. Boschi and A-H. A. Park (2014c). Morphological changes during enhanced carbonation of asbestos containing material and its comparison to magnesium silicate minerals. *Journal of Hazardous Materials* 264: 42–52.

Geerlings, H. and R. Zevenhoven (2013). CO₂ mineralization — bridge between storage and utilization of CO₂. *Annual Review of Chemical and Biomolecular Engineering* 4: 103–117.

Gerdemann, S. J., W. K. O'Connor, D. C. Dahlin, L. R. Penner and H. Rush (2007). *Ex-Situ* Aqueous Mineral Carbonation. *Environmental Science & Technology* 41(7): 2587–2593.

Ghoorah, M., B. Z. Dlugogorski, R. D. Balucan and E. M. Kennedy (2014). Selection of acid for weak acid processing of wollastonite for mineralisation of CO₂. *Fuel* 122: 277–286.

Giannoulakis, S., K. Volkart and C. Bauer (2014). Life cycle and cost assessment of mineral carbonation for carbon capture and storage in European power generation. *International Journal of Greenhouse Gas Control* 21: 140–157.

GMD (2014) last update, Up-to-date weekly average CO₂ at Mauna Loa. Retrieved 10/06/2016. Available from http://www.esrl.noaa.gov/gmd/ccgg/trends/weekly.html.

Goff, F. and K. S. Lackner (1998). Carbon dioxide sequestering using ultramafic rocks. *Environmental Geosciences* 5(3): 89–101.

Halstead, W. D. (1970). Thermal decomposition of ammonium sulphate. *Journal of Applied Chemistry* 20(4): 129–132.

Hänchen, M., S. Krevor, M. Mazzotti and K. S. Lackner (2007). Validation of a population balance model for olivine dissolution. *Chemical Engineering Science* 62(22): 6412–6422.

Hänchen, M., V. Prigiobbe, R. Baciocchi and M. Mazzotti (2008). Precipitation in the Mg-carbonate system — effects of temperature and CO₂ pressure. *Chemical Engineering Science* 63(4): 1012–1028.

Hariharan, S. B., M. Werner, D. Zingaretti, R. Baciocchi and M. Mazzotti (2013). Dissolution of activated serpentine for direct flue-gas mineralization. *Energy Procedia* 37: 5938–5944.

Hariharan, S., M. Werner, M. Hänchen and M. Mazzotti (2014). Dissolution of dehydroxylated lizardite at flue gas conditions: II. Kinetic modeling. *Chemical Engineering Journal* 241: 314–326.

Hariharan, S., M. Werner and M. Mazzotti (2015). Single and multi-strep processes for flue gas CO₂ mineralisation using thermally activated serpentine. Presented at ACEME2015, New York City, NY, USA, June 2015, 7 pages.

Haug, T. A., R. A. Kleiv and I. A. Munz (2010). Investigating dissolution of mechanically activated olivine for carbonation purposes. *Applied Geochemistry* 25(10): 1547–1563.

Hemmati, A., J. Shayegan, J. Bu, T. Y. Yeo and P. Sharratt (2014a). Process optimization for mineral carbonation in aqueous phase. *International Journal of Mineral Processing* 130: 20–27.

Hemmati, A., J. Shayegan, P. Sharratt, T. Y. Yeo and J. Bu (2014b). Solid products characterization in a multi-step mineralization process. *Chemical Engineering Journal* 252: 210–219.

Highfield, J., J. Chen, J. Bu, J. Åbacka, J. Fagerlund and R. Zevenhoven (2013). Steam-promoted gas–solid carbonation of magnesia and brucite below 200° C, *Proceedings of ACEME13*, Leuven, Belgium, April 10–12: 161–172.

Highfield, J., H. Lim, J. Fagerlund and R. Zevenhoven (2012). Activation of serpentine for CO₂ mineralization by flux extraction of soluble magnesium salts using ammonium sulfate. *RSC Advances* 2: 6535–6541.

Highfield, J. G., H. Q. Lim, J. Fagerlund and R. Zevenhoven (2012b). Mechanochemical activation of serpentine for CO₂ mineralization using ammonium salts under ambient conditions. *RSC Advances* 2(2012): 6542–6548.

Highfield, J., J. Åbacka, J., Chen, E. Nduagu and R. Zevenhoven (2015). An overview of ÅAU/ICES cooperation in *ex-situ* CO₂ mineralization. Presented (as poster) at ICCDU2015, Singapore.

Hill, R. J., J. H. Canterford and F. J. Moyle (1982). New data for Lansfordite. *Mineralogical Magazine* 46: 453–457.

House, K. Z., C. H. House, D. P. Schrag and M. J. Aziz (2007). Electro chemical acceleration of chemical weathering as an energetically feasible approach to mitigating anthropogenic climate change. *Environmental Science & Technology* 41(24): 8464–8470.

Huijgen, W. J. J. and R. N. J. Comans (2003). Carbon dioxide sequestration by mineral carbonation. Literature Review. ECN-C--03-016. Petten, the Netherlands: Energy Research Center of the Netherlands.

Huijgen, W. J. J. and R. N. J. Comans (2005). Carbon dioxide sequestration by mineral carbonation, Literature review update 2003-2004. ECN-C--05--022. Petten, the Netherlands: Energy Research Center of the Netherlands.

Huijgen, W. J. J., R. N. J. Comans and G. Witkamp (2007). Cost evaluation of CO₂ sequestration by aqueous mineral carbonation. *Energy Conversion and Management* 48(7): 1923–1935.

IEA GHG (2013). Mineralisation — carbonation and enhanced weathering. Available from: http://ieaghg.org/docs/General_Docs/Reports/2013-TR6.pdf.

IEA GHG (2015). Integrated Carbon Capture and Storage Project at SaskPower's Boundary Dam Power Station. Available from: http://www.ieaghg.org/docs/General_Docs/Reports/2015-06.pdf. Retrieved 10/06/2016.

IPCC (2005). *IPCC Special Report on Carbon Dioxide Capture and Storage.* Prepared by Working Group III of the Intergovernmental Panel on Climate Change [Metz, B., O. Davidson, H. C. de Coninck, M. Loos and L. A. Meyer, (eds.)]. Cambridge University Press, Cambridge, United Kingdom and New York, NY, USA, p. 442.

IPCC (2007). Summary for policymakers. In: *Climate Change 2007: The Physical Science Basis* [Solomon, S., D. Qin, M. Manning, Z. Chen, M. Marquis, K. B. Averyt, M. Tignorand and H. L. Miller's (eds.)]. Contribution of Working Group I to the Fourth Assessment Report of the Intergovernmental Panel on Climate Change. Available from: https://www.ipcc.ch/publications_and_data/publications_ipcc_fourth_assessment_report_wg1_report_the_physical_science_basis.htm. Cambridge University Press, Cambridge, United Kingdom and New York, NY, USA.

IPCC (2013). Summary for Policymakers. In: *Climate Change 2013: The Physical Science Basis.* Contribution of Working Group I to the Fifth Assessment Report of the Intergovernmental Panel on Climate Change [Stocker, T. F., D. Qin, G.-K. Plattner, M. Tignor, S. K. Allen, J. Boschung, A. Nauels, Y. Xia, V. Bex and P. M. Midgley (eds.)]. Cambridge University Press, Cambridge, United Kingdom and New York, NY, USA.

IPCC (2014). *Climate Change 2014: Synthesis Report.* Contribution of Working Groups I, II and III to the Fifth Assessment Report of the Intergovernmental Panel on Climate Change [Core Writing Team, Pachauri, R. K. and L. A. Meyer, (eds.)] IPCC, Geneva, Switzerland, p. 151.

Julcour, C., F. Bourgeois, B. Bonfils, I. Benhamed, F. Guyot, F. Bodénan, C. Petiot and É. C. Gaucher (2015). Development of an attrition-leaching hybrid process for direct aqueous mineral carbonation. *Chemical Engineering Journal* 262: 716–726.

Kakizawa, M., A. Yamasaki and Y. Yanagisawa (2001). A new CO₂ disposal process using artificial rock weathering of calcium silicate accelerated by acetic acid. *Energy* 26: 341–354.

Khalilpour, R., K. Mumford, H. Zhai, A. Abbas, G. Stevens and E. S. Rubin (2015). Membrane-based carbon capture from flue gas: A review. *Journal of Cleaner Production* 103: 286–300.

Khoo, H. H., P. N. Sharatt, J. Bu, A. Borgna, T. Y. Yeo, J. Highfield, T. G. Björklöf and R. Zevenhoven (2011a). Carbon capture and mineralization in Singapore: preliminary environmental impacts and costs via LCA. *Industrial & Engineering Chemistry Research* 50: 11350–11357.

Khoo, H. H., J. Bu, R. L. Wong, S. Y. Kuan and P. N. Sharratt (2011b). Carbon capture and utilization: Preliminary life cycle CO₂, energy, and cost results of potential mineral carbonation. *Energy Procedia* 4: 2494–2501.

Kirchofer, A., A. Brandt, S. Krevor, V. Priogiobbe and J. Wilcox (2012). Impact of alkalinity sources on the life-cycle energy efficiency of mineral carbonation technologies. *Energy & Environmental Science* 5: 8631–8641.

Kiyoura, R. and K. Urano (1970). Mechanism, Kinetics, and Equilibrium of Thermal Decomposition of Ammonium Sulfate. *Industrial & Engineering Chemistry Process Design and Development* 9(4): 489–494.

Knudsen J., P.-J. Vilhemsen, J. Jensen and O. Biede (2007). First year operating experience with a 1 t/h CO_2 absorption Pilot plant at Esbjerg coal-fired power plant. *VGB Power Tech* 3: 57–61.

Kohlmann, J. and R. Zevenhoven (2001). The removal of CO_2 from flue gases using magnesium silicates in Finland. *11th International Conference on Coal Science (ICCS-11)*, San Francisco, CA, September 2001.

Koivisto, E. (2013). *Utilization Potential of Iron Oxide by Product from Serpentinite Carbonation.* Master of Science in Engineering Technology Thesis Luleå University of Technology, Sweden.

Koljonen, T., L. Similä, K. Sipilä, S. Helynen, M. Airaksinen, J. Laurikko, J. Manninen, T. Mäkinen, A. Lehtilä, J. Honkatukia, P. Tuominen, T. Vainio, T. Järvi, K. Mäkelä, S. Vuori, J. Kiviluoma, K. Sipilä, J. Kohl, M. Nieminen (2012). *Low Carbon Finland 2050. VTT Clean Energy Technology Strategies for Society.* VTT, Espoo, Finland. 75 pages.

Krevor, S. C. and K. S. Lackner (2009). Enhancing process kinetics for mineral carbon sequestration. *Energy Procedia* 1(1): 4867–4871.

Krevor, S. C. M. and K. S. Lackner (2011). Enhancing serpentine dissolution kinetics for mineral carbon dioxide sequestration. *International Journal of Greenhouse Gas Control* 5(4): 1073–1080.

Kwon, S. (2011). PhD thesis: *Mineralization for CO_2 Sequestration Using Olivine Sorbent in the Presence of Water Vapor.* Georgia Institute of Technology. Atlanta, GA, USA.

Kwon, S., M. Fan, H. F. M. DaCosta, A. G. Russell (2011). Factors affecting the direct mineralization of CO_2 with olivine. *Journal of Environmental Sciences* 23(8): 1233–1239.

Lackner, K. S. (2003). A guide to CO_2 sequestration. *Science* 300(5626): 1677–1678.

Lackner, K. S., C. H. Wendt, D. P. Butt, E. L. Joyce and D. H. Sharp (1995). Carbon dioxide disposal in carbonate minerals. *Energy* 20: 1153–1170.

Lackner, K. S., D. P. Butt and C. H. Wendt (1997). Progress on binding CO_2 in mineral substrates. *Energy Conversion and Management* 38 (Supplement): S259–S264.

Larachi, F., J. Gravel, B. P. A. Grandjean and G. Beaudoin (2012). Role of steam, hydrogen and pretreatment in chrysotile gas–solid carbonation: Opportunities for pre-combustion CO_2 capture. *International Journal of Greenhouse Gas Control* 6: 69–76.

Li, W., W. Li, B. Li and Z. Bai (2009). Electrolysis and heat pretreatment methods to promote CO_2 sequestration by mineral carbonation. *Chemical Engineering Research & Design* 87: 210–215.

Li, L., N. Zhao, W. Wei and Y. Sun (2013). A review of research progress on CO_2 capture, storage, and utilization in Chinese Academy of Sciences. *Fuel* 108: 112–130.

Li, T., T. C. Keener and L. Cheng (2014). Carbon dioxide removal by using $Mg(OH)_2$ in a bubble column: Effects of various operating parameters. *International Journal of Greenhouse Gas Control* 31: 67–76.

Lockwood, M. and C. Fröhlich (2008). Recent oppositely directed trends in solar climate forcings and the global mean surface air temperature. II. Different reconstructions of the total solar irradiance variation and dependence on response time scale. *Proceedings of the Royal Society A: Mathematical, Physical and Engineering Science* 464(2094): 1367–1385.

Lupion, M., I. Alvarez, P. Otero, R. Kuivalainen, J. Lantto, A. Hotta and H. Hack (2013). 30 MWth CIUDEN Oxy-cfb Boiler — First Experiences. *Energy Procedia* 37: 6179–6188.

Madeddu, S., M. Priestnall, H. Kinoshita and E. Godoy (2014). Alkaline digestion of dunite for Mg(OH)2 production: An investigation for indirect CO_2 sequestration. *Minerals Engineering* 59: 31–38.

Maroto-Valer, M., Y. Zhang, M. E. Kuchta, J. M. Andrésen and D. J. Fauth (2005). Process for sequestering carbon dioxide and sulphur dioxide. US Patent US2005/0002847.

Mattila, H. and R. Zevenhoven (2014). Chapter Ten — Production of Precipitated Calcium Carbonate from Steel Converter Slag and Other Calcium-Containing Industrial Wastes and Residues. *Advances in Inorganic Chemistry* 66: 347–384.

Mazari, S. A., B. S. Ali, B. M. Jan, I. N. Saeed and S. Nizamuddin (2015). An overview of solvent management and emissions of amine-based CO_2 capture technology. *International Journal of Greenhouse Gas Control* 34: 129–140.

McKelvy, M. J., A. V. G. Chizmeshya, J. Diefenbacher, H. Béarat and G. Wolf (2004). Exploration of the Role of Heat Activation in Enhancing Serpentine Carbon Sequestration Reactions. *Environmental Science & Technology* 38(24): 6897–6903.

Mehleri, E. D., A. Bhave, N. Shah, P. Fenneli and M. C. Dowell (2015). Techno-economic assessment and environmental impacts of mineral carbonation of industrial wastes and other uses of carbon dioxide Presented at ACEME2015. New York City, NY, USA, June 2015, 10 pages.

Meyer, N. A., J. U. Vögeli, M. Becker, J. L. Broadhurst, D. L. Reid and J.-P. Franzidis (2014). Mineral carbonation of PGM mine tailings for CO_2 storage in South Africa: A case study. *Minerals Engineering* 59: 45–51.

Moosdorf, N., P. Renforth and J. Hartmann (2014). Carbon dioxide efficiency of terrestrial enhanced weathering. *Environmental Science & Technology* 48(9): 4809–4816.

Mulargia, F. and A. Bizzarri (2014). Anthropogenic triggering of large earthquakes. *Scientific Reports* 4(article): 6100.

Nduagu, E. (2008). Mineral carbonation: Preparation of magnesium hydroxide, Mg(OH)2, from serpentinite rock; Nduagu, E. (2012). Dr. Tech. Thesis. *Production of Mg(OH)2 from Mg-silicate rock for CO2 Mineral Sequestration.* Available from: https://www.doria.fi/bitstream/handle/10024/86170/nduagu%20experience.pdf?sequence=2.

Nduagu, E., T. Björklöf, J. Fagerlund, J. Wärnå, H. Geerlings and R. Zevenhoven (2012a). Production of magnesium hydroxide from magnesium silicate for the purpose of CO_2 mineralisation — Part 1: Application to Finnish serpentinite. *Minerals Engineering* 30: 75–86.

Nduagu, E., T. Björklöf, J. Fagerlund, E. Mäkilä, J. Salonen, H. Geerlings and R. Zevenhoven (2012b). Production of magnesium hydroxide from magnesium silicate for the purpose of CO_2 mineralization — Part 2: Mg extraction modeling and application to different Mg silicate rocks. *Minerals Engineering* 30: 87–94.

Nduagu, E., J. Bergerson and R. Zevenhoven (2012c). Life cycle assessment of CO_2 sequestration in magnesium silicate rock — a comparative study. *Energy Conversion and Management* 55: 116–126.

Nduagu, E., I. Romão, J. Fagerlund and R. Zevenhoven (2013). Performance assessment of producing $Mg(OH)_2$ for CO_2 mineral sequestration. *Applied Energy* 106: 116–126.

Nduagu, E. I., J. Highfield, J. Chen and R. Zevenhoven (2014). Mechanisms of serpentine-ammonium sulfate reactions studied by coupled thermal-spectroscopic methods: Towards higher efficiencies in flux recovery and Mg extraction for CO_2 mineral sequestration. *RSC Advances* 4: 64494–64505.

Newall, P. S., S. J. Clarke, H. M. Haywood, H. Scholes, N. R. Clarke, P. A. King and R. W. Barley (1999). CO_2 storage as carbonates minerals. PH 3/17. IEA Greenhouse Gas R&D Programme. C. C. Ltd. Cornwall, UK.

O'Connor, W. K., D. C. Dahlin, D. N. Nilsen, S. J. Gerdemann, G. E. L. Rush, R. Penner, R. P. Walters and P. C. Turner (2002). Continuing studies on direct aqueous mineral carbonation for CO_2 sequestration. *The Proceedings of the 27th International Technical Conference on Coal Utilization & Fuel Systems*, pp. 819–830.

O'Connor, W. K., D. C. Dahlin, G. E. Rush, S. J. Gerdemann, L. R. Penner and D. N. Nilsen (2005). Final report — Aqueous mineral carbonation. Mineral availability, pre-treatment, reaction parametrics and process studies. DOE/ARC-TR-04-002.

Olajire, A. A. (2013). A review of mineral carbonation technology in sequestration of CO_2. *Journal of Petroleum Science and Engineering* 109: 364–392.

Oliver, J. G. J., G. Janssens-Maenhout, M. Muntean and J. A. H. W. Peters (2013). Trends in global CO_2 emissions. The Hague: PBL Netherlands Environmental Assessment Agency Ispra Joint Research Centre. ISBN: 978-94-91506-51-2.

Olsen, A. A. and J. Donald Rimstidt (2008). Oxalate-promoted forsterite dissolution at low pH. *Geochimica et Cosmochimica Acta* 72(7): 1758–1766.

Park, A. A., R. Jadhav and L. Fan (2003). CO_2 Mineral sequestration: Chemically enhanced aqueous carbonation of serpentine. *The Canadian Journal of Chemical Engineering* 81(3–4): 885–890.

Park, Y., K.-Y. A. Lin, A.-H. A. Park and C. Petit (2015). Recent Advances in Anhydrous Solvents for CO_2 Capture: Ionic Liquids, Switchable Solvents, and Nanoparticle Organic Hybrid Materials. Frontiers in Energy Research 3, article 42.

Pasquier, L., G. Mercier, J. Blais, E. Cecchi and S. Kentish (2014a). Reaction mechanism for the aqueous-phase mineral carbonation of heat-activated serpentine at low temperatures and pressures in flue gas conditions. *Environmental Science & Technology* 48(9): 5163–5170.

Pasquier, L., G. Mercier, J. Blais, E. Cecchi and S. Kentish (2014b). Parameters optimization for direct flue gas CO_2 capture and sequestration by aqueous mineral carbonation using activated serpentinite based mining residue. *Applied Geochemistry* 50: 66–73.

Penner, L., W. K. O'Connor, D. C. Dahlin, S. Gerdemann and G. E. Rush (2004). Mineral carbonation: Energy costs of pretreatment options and insights gained from flow loop reaction studies. *Proceedings of the Third Annual Conference on Carbon Capture & Sequestration* Alexandria, VA, USA, pp. 3–6.

Pereira, N., J. F. Carneiro, A. Araújo, M. Bezzeghoud and J. Borges (2014). Seismic and structural geology constraints to the selection of CO_2 storage sites — The case of the onshore Lusitanian basin, Portugal. *Journal of Applied Geophysics* 102: 21–38.

Perrin, N., R. Dubettier, F. Lockwood, J. Tranier, C. Bourhy-Weber and P. Terrien (2015). Oxycombustion for coal power plants: Advantages, solutions and projects. *Applied Thermal Engineering* 74: 75–82.

Power, I. M., A. L. Harrison, G. M. Dipple, S. A. Wilson, P. Kelemen, M. Hitch and G. Southam (2013). Carbon mineralization: From natural analogues to engineered systems. *Reviews in Mineralogy and Geochemistry* 77: 305–360.

Prigiobbe, V., M. Hänchen, G. Costa, R. Baciocchi and M. Mazzotti (2009a). Analysis of the effect of temperature, pH, CO_2 pressure and salinity on the olivine dissolution kinetics. *Energy Procedia* 1(1): 4881–4884.

Prigiobbe, V., M. Hänchen, M. Werner, R. Baciocchi and M. Mazzotti (2009b). Mineral carbonation process for CO_2 sequestration. *Energy Procedia* 1(1): 4885–4890.

Rimstidt, J. D., S. L. Brantley and A. A. Olsen (2012). Systematic review of forsterite dissolution rate data. *Geochimica et Cosmochimica Acta* 99: 159–178.

Romão, I. S. (2015). Dr. Tech Thesis. *Production of Magnesium Carbonates from Serpentinites for CO_2 Mineral Sequestration — Optimisation Towards Industrial Application*. Åbo Akademi University, Åbo/Turku, Finland & University of Coimbra, Portugal.

Romão, I., M. Eriksson, E. Nduagu, J. Fagerlund, L. M. Gando-Ferreira, R. Zevenhoven (2012a). Carbon dioxide storage by mineralisation applied to an industrial-scale lime kiln. *ECOS 2012 Proceedings of the 25th International Conference on Efficiency, Cost, Optimization, Simulation and Environmental Impact of Energy Systems*; 2012 June 26–29. Perugia, Italy, paper 226.

Romão, I., E. Nduagu, J. Fagerlund, L. M. Gando-Ferreira and R. Zevenhoven (2012b). CO_2 Fixation Using Magnesium Silicate Minerals. Part 2: Energy Efficiency and Integration with Iron-and Steelmaking. *Energy* 41: 203–211.

Romão, I. S. and I. Grigaliûnaitë (2013). Water and aqueous solution processing for mineral carbonation. unpublished report, Åbo Akademi University, Åbo/Turku Finland. p. 29.

Romão, I., M. Slotte, L. M. Gando-Ferreira and R. Zevenhoven (2014). CO_2 sequestration with Magnesium Silicates — Exergetic performance assessment. *Chemical Engineering Research and Design* 92: 2072–2082.

Romão, I. S., L. M. Gando Ferreira, M. M. V. G. da Silva and R. Zevenhoven (2016). CO_2 sequestration with serpentinite and metaperodotite from North-East Portugal. *Minerals Engineering* 94: 101–114.

Rozalen, M. and F. J. Huertas (2013). Comparative effect of chrysotile leaching in nitric, sulfuric and oxalic acids at room temperature. *Chemical Geology* 352: 134–142.

Rubin, E. S., C. Chen and A. B. Rao (2007). Cost and performance of fossil fuel power plants with CO_2 capture and storage. *Energy Policy* 25: 4444–4454.

Salek, S. S., R. Kleerebezem, H. M. Jonkers, G. J. Witkamp and M. C. M. van Loosdrecht (2013). Mineral CO_2 sequestration by environmental biotechnological processes. *Trends in Biotechnology* 31: 139–146.

Sanna, A. and M. Maroto-Valer (2014). CO_2 Sequestration Using a Novel Na-salts pH Swing Mineral Carbonation Process. *Energy Procedia* 63: 5897–5903.

Sanna, A., M. Dri and M. Maroto-Valer (2013). Carbon dioxide capture and storage by pH swing aqueous mineralisation using a mixture of ammonium salts and antigorite source. *Fuel* 114: 153–161.

Sanna, A., M. Uibu, G. Caramanna, R. Kuusik and M. Maroto-Valer (2014). A review of mineral carbonation technologies to sequester CO_2. *Chemical Society Reviews* 43: 8049–8080.

Schaetzl, R. and M. L. Thompson (eds.) (2015). *Soils Genesis and Geomorphology*. 2nd edn, Cambridge University Press, New York.

Seifritz, W. (1990). CO_2 disposal by means of silicates. *Nature* 345: 486.

Seixas, J., P. Fortes, L. Dias, J. Carneiro, P. Mesquita, D. Boavida, R. Aguiar, F. Marques, V. Fernandes, J. Helseth, J. Ciesielska and K. Whiriskey (2015). CO_2 Capture and Storage in Portugal. A bridge to a low carbon economy. Universidade Nova de Lisboa. Faculdade de Ciências e Tecnologia, Lisboa, 2015.

Sipilä, J., S. Teir and R. Zevenhoven (2008). Carbon dioxide sequestration by mineral carbonation. Literature review update 2005–2007. VT 2008-1. Åbo Akademi University, Turku, Finland.

Slotte, M., I. Romão and R. Zevenhoven (2013). Challenges in process scale-up of serpentinite carbonation to pilot scale. *Energy* 62: 142–149.

Slotte, M. and R. Zevenhoven (2013). Total lime kiln gas compression for CO_2 mineral sequestration, ECOS'2013. *26th International Conference on Efficiency, Cost, Optimization, Simulation and Environmental Impact of Energy Systems*, July 15–19, 2013, Guilin, China, paper F005.

Stasiulaitiene, I. J. Fagerlund, E. Nduagu, G. Denafas and R. Zevenhoven (2011). Carbonation of serpentinite rock from Lithuania and Finland. *Energy Procedia (GHGT-10)* 4: 2963–2970.

Stasiulaitiene, I., A. Babarskaite, D. Martuzevicius and R. Zevenhoven (2013). Comparison of mineral carbonation process with geological carbon dioxide storage by life cycle assessment tool. *Proceedings of ACEME 2013*. KU Leuven, Belgium, April 9–12, 2013.

Stasiulaitiene, I., V. Vajegaite, D. Martuzevicius, G. Denafas, S. Sliaupa, J. Fagerlund and R. Zevenhoven (2014). Parameters affecting $Mg(OH)_2$ extraction from serpentinites in Lithuania for the purpose of CO_2 reduction by mineral carbonation. *Environmental Progress & Sustainable Energy* 33(2): 512–518.

Steel, K. M. and R. D. Balucan (2015). Recent advances in the development of a new pH swing method based on a regenerable precipitant-solvent system for metals recovery. Presented at ACEME2015. New York City, NY, USA, June 2015, p. 7.

Stephen A. R. (2010). *Carbon Capture and Storage*. Oxford, Butterworth-Heinemann.

Styring, P., H. de Coninck and K. Armstrong (2011). *Carbon Capture and Utilisation in the Green Economy*. CO_2 Chem Media & Publishing. Centre for low carbon futures. York (UK).

Swanson, E. J., K. J. Fricker, M. Sun and A.-H. A. Park (2014). Directed precipitation of hydrated and anhydrous magnesium carbonates for carbon storage. *Physical Chemistry Chemical Physics* 16(42): 23440–23450.

Szargut, J., D. R. Morris and F. R. Steward (1988). *Exergy Analysis of Thermal, Chemical, and Metallurgical Processes*. Hemisphere, New York.

Teir, S. (2008). Dr. thesis. *Fixation of Carbon Dioxide by Producing Carbonates from Minerals and Steelmaking Slags*. Helsinki University of Technology, Espoo, Finland.

Teir, S., E. Tsupari, A. Arasto, T. Koljonen, J. Kärki, A. Lehtilä, L. Kujanpää, S. Aatos and M. Nieminen (2011). Prospects for application of CCS in Finland. *Energy Procedia* 4: 6174–6181.

Teir, S., S. Eloneva, S. Aatos, O.-P. Isomäki, C.-J. Fogelholm and R. Zevenhoven (2006a). Carbonation of Finnish magnesium silicates for CO_2 sequestration. Presented (as poster) at *The Fifth Annual Conference on Carbon Capture & Sequestration*, Alexandria (VA).

Teir, S., S. Eloneva, C. Fogelholm and R. Zevenhoven (2006b). Stability of calcium carbonate and magnesium carbonate in rainwater and nitric acid solutions. *Energy Conversion and Management* 47(18–19): 3059–3068.

Teir, S., R. Kuusik, C. Fogelholm and R. Zevenhoven (2007a). Production of magnesium carbonates from serpentinite for long-term storage of CO_2. *International Journal of Mineral Processing* 85(1–3): 1–15.

Teir, S., H. Revitzer, S. Eloneva, C. Fogelholm and R. Zevenhoven (2007b). Dissolution of natural serpentinite in mineral and organic acids. *International Journal of Mineral Processing* 83(1–2): 36–46.

Teir, S., S. Eloneva, C.-J. Fogelholm and R. Zevenhoven (2009). Fixation of carbon dioxide by producing hydromagnesite from serpentinite. *Applied Energy* 86: 214–218.

Torróntegui, M. D. (2010). MSc Thesis. *Assessing the Mineral Carbonation Science and Technology*. ETH-Zurich, Swiss Federal Institute of Technology.

Trenberth, K. E., J. T. Fasullo and J. Kiehl (2009). Earth's global energy budget. *Bulletin of the American Meteorological Society* 90: 311–323.

Turianicová, E., P. Baláž, L. Tuček, A. Zorkovská, V. Zeleòák, Z. Németh, A. Šatka and J. Kováè (2013). A comparison of the reactivity of activated and non-activated olivine with CO₂. *International Journal of Mineral Processing* 123: 73–77.

Veetil, S. P., G. Mercier, J. Blais, E. Cecchi and S. Kentish (2015). Magnetic separation of serpentinite mining residue as a precursor to mineral carbonation. *International Journal of Mineral Processing* 140: 19–25.

Verduyn, M., H. Boerrigter, R. Oudwater and G. A. F. van Mossel (2009). A novel process concept for CO₂ mineralization; technical opportunities and challenges. Presented at *The Fifth Trondheim Conference on CO₂ Capture, Transport and Storage* (TCCS-5 16–17 June 2009, Trondheim Norway).

Virtanen, M. (2015) MSc thesis. *Energy-Efficient Solutions for Concentrating or Separating Dissolved Ammonium Sulphate*. Åbo Akademi University, Åbo/Turku Finland.

Wang, X. and M. Maroto-Valer (2011a). Dissolution of serpentine using recyclable ammonium salts for CO₂ mineral carbonation. *Fuel* 90(3): 1229–1237.

Wang, X. and M. Maroto-Valer (2011b). Integration of CO₂ capture and storage based on pH-swing mineral carbonation using recyclable ammonium salts. *Energy Procedia* 4: 4930–4936.

Wang, X. and M. Maroto-Valer (2013). Optimization of carbon dioxide capture and storage with mineralisation using recyclable ammonium salts. *Energy* 51: 431–438.

Wang, X., M. Maroto-Valer, G. Shiwang and X. Shisen (2013). Preliminary cost evaluation of integrated aqueous ammonia capture with mineralisation using recyclable salts for distributed CCS. *Energy Procedia* 37: 2529–2535.

Wendt, C. H., D. P. Butt, K. S. Lackner and H. Ziock (1998). Thermodynamic considerations of using chlorides to accelerate the carbonate formation from magnesium silicates, LA-UR-98-3612. Los Alamos National Laboratory, Los Alamos, NM, USA.

Werner, M., S. B. Hariharan, A. V. Bortolan, D. Zingaretti, R. Baciocchi and M. Mazzotti (2013). Carbonation of Activated Serpentine for Direct Flue Gas Mineralization. *Energy Procedia* 37: 5929–5937.

Werner, M., S. Hariharan, D. Zingaretti, R. Baciocchi and M. Mazzotti (2014). Dissolution of dehydroxylated lizardite at flue gas conditions: I. Experimental study. *Chemical Engineering Journal* 241: 301–313.

Yeo, T. Y., A. Hemmati, P. Sharratt and J. Bu (2015). Effect of particle sizes on mineralization process efficiency. Presented at *The 13th International Conference on CO₂ Utilisation, ICCDU2015*, Singapore, 5–9 July 2015.

Zevenhoven, R. and J. Kohlmann (2002). CO₂ sequestration by magnesium silicate mineral carbonation in Finland. Recovery, Recycling & Re-integration, Geneva, Switzerland. 2002, paper 220.

90   *CO₂ Sequestration by Ex-Situ Mineral Carbonation*

Zevenhoven, R., S. Eloneva and S. Teir (2006). Chemical fixation of CO_2 in carbonates: Routes to valuable products and long-term storage. *Catalysis Today* 115: 73–79.

Zevenhoven, R., S. Teir and S. Eloneva (2008). Heat optimisation of a staged gas–solid mineral carbonation process for long-term CO_2 storage. *Energy* 33(2): 362–370.

Zevenhoven, R., J. Sipilä and S. Teir (2008). Motivations for carbonating magnesium silicates using a gas-solid process route. *Proceedings of ACEME2008*, Rome, Italy, October 1–3, 2008, paper 016, p. 10.

Zevenhoven, R. and J. Fagerlund (2010). Mineralisation of CO_2. In: *Chapter 16 Developments and Innovation CCS Technology* [Maroto-Valer's, M. (ed.)]. Woodhead Publishing Ltd., Cambridge, UK, pp. 433–462.

Zevenhoven, R., J. Fagerlund, T. Björklöf, M. Mäkelä and O. Eklund (2012). Carbon dioxide mineralisation and integration with flue gas desulphurisation applied to a modern coal-fired power plant. *Proceedings of ECOS'2012*, Perugia, Italy, June 2012, paper 179, p. 20.

Zevenhoven, R., J. Fagerlund, E. Nduagu, I. Romão, B. Jie and J. Highfield (2013). Carbon storage by mineralisation (CSM): Serpentinite rock carbonation via $Mg(OH)_2$ reaction intermediate without CO_2 pre-separation. *Energy Procedia (GHGT-11)* 37: 5945–5954.

Zevenhoven, R., M. Slotte, J. Åbacka and J. Highfield (2015). A comparison of CO_2 mineral carbonation processes involving a dry or wet carbonation step, Proceedings of ECOS'2015. *28th International Conference on Efficiency, Cost, Optimization, Simulation and Environmental Impact of Energy Systems*, June 30–July 3, Pau, France, paper 51128, p. 12.

Zhang, Q., K. Sugiyama and F. Saito (1996). Enhancement of acid extraction of magnesium and silicon from serpentine by mechanochemical treatment. *Hydrometallurgy* 45: 323–331.

Zhao, L., L. Sang, J. Chen, J. Ji and H. H. Teng (2010). Aqueous Carbonation of Natural Brucite: Relevance to CO_2 Sequestration. *Environmental Science & Technology* 44(1): 406–411.

Chapter 4

MC Technologies Developed
for Waste Materials

Mai Uibu, Regiina Viires and Rein Kuusik
*Laboratory of Inorganic Materials, Tallinn University of Technology,
Ehitajate tee 5, 19086 Tallinn, Estonia*

4.1 Introduction

In addition to primary earth minerals, the solid wastes from industrial processes including fossil fuel power station, solid waste incinerator, iron- and steel plant, cement plant, mining and, paper industry (Figure. 4.1) could also be used as a feedstock for mineral carbonation (MC) (Wee, 2013; Eloneva *et al.*, 2008b). This method has several advantages, as these materials are mostly accompanied by CO_2 point source emissions, providing obtainable calcium and magnesium source in the form of free oxides and/or silicates without the necessity for mining. Inorganic wastes are typically fine-grained and appear to be chemically more unstable compared to natural earth minerals (Huijgen and Comans, 2006). Due to that, they need less pre-processing and milder operating conditions to achieve the same level of carbonation yields (Huijgen and Comans, 2003, 2006).

Secondary benefits of MC include improved environmental quality through mineral transformation and neutralisation (municipal solid waste incinerator (MSWI) residues, air pollution control (APC) wastes, asbestos tailings, red mud (RM), oil shale fly ash (FA)) as

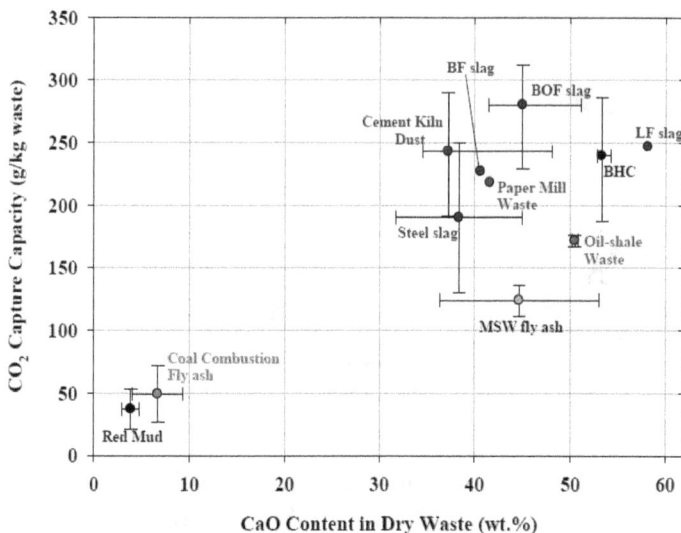

Figure 4.1 CO_2 capture capacity of different kinds of solid wastes (adapted from Pan *et al.*, 2012).

well as utilisation of the end-products such as construction materials (Huntzinger *et al.*, 2009b) or PCC-based fillers (Velts *et al.*, 2011; Eloneva *et al.*, 2008b). However, the amounts available are rather limited and unpredictable as the technology (differences in chemical composition and obtainability) and legislation changes (Sanna *et al.*, 2012a).

4.2 Process Chemistry and Main Carbonation Routes

The most common MC route for alkaline wastes is the direct aqueous route (DAC), in which the dissolution of reactive phases and the precipitation of Ca(Mg)-carbonates occur in a single stage. Direct aqueous carbonation is usually performed either in slurry phase at liquid to solid (L/S) of 5–50 w/w (Back *et al.*, 2008; Bonenfant *et al.*, 2008a; Uibu *et al.*, 2009; Huijgen *et al.*, 2005) or via the thin-film route — L/S ratios <1.5 w/w (Baciocchi *et al.*, 2009a, 2015; Uibu *et al.*, 2009; Gunning *et al.*, 2010; Huntzinger *et al.*, 2009a, Bertos *et al.*, 2004b).

The process chemistry corresponds to the type and composition of waste material. Industrial wastes from solid fuel combustion (lignite FA, oil shale (OS) FA) and from MSWI APC often contain a substantial amount of free CaO as the main CO_2-binding compound. MC is in that case usually carried out in aqueous conditions according to Reactions (4.1)–(4.5): irreversible CaO hydration (Reaction (4.1)) is followed by concurrent dissolution of $Ca(OH)_2$ (Reaction (4.2)) and dissociation of aqueous CO_2 (Reactions (4.3), (4.4)), until the Ca^{2+}–ions are converted to $CaCO_3$ and precipitated out (Reaction (4.5)) (Pan *et al.*, 2012; Montes-Hernandez *et al.*, 2009; Uibu *et al.*, 2010).

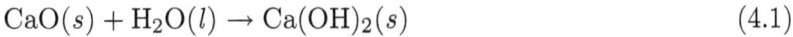

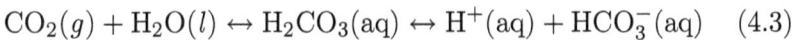

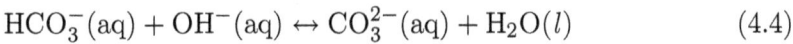

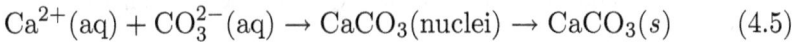

$$CaO(s) + H_2O(l) \rightarrow Ca(OH)_2(s) \tag{4.1}$$

$$Ca(OH)_2(s) \xrightarrow{\text{water}} Ca^{2+}(aq) + 2OH^-(aq) \tag{4.2}$$

$$CO_2(g) + H_2O(l) \leftrightarrow H_2CO_3(aq) \leftrightarrow H^+(aq) + HCO_3^-(aq) \tag{4.3}$$

$$HCO_3^-(aq) + OH^-(aq) \leftrightarrow CO_3^{2-}(aq) + H_2O(l) \tag{4.4}$$

$$Ca^{2+}(aq) + CO_3^{2-}(aq) \rightarrow CaCO_3(\text{nuclei}) \rightarrow CaCO_3(s) \tag{4.5}$$

MC of wastes that contain Ca-silicates as the main CO_2-binders (such as steel slags, MSWI bottom ash (BA), etc.) proceeds less rapidly: The kinetics of CO_2 uptake include subsequent reaction stages: (1) a rapid CO_2 uptake with time, which involves $Ca(OH)_2$ (Reactions (4.1)–(4.5)) and (2) a decline in the rate until stabilisation of CO_2 uptake at constant value (Rendek *et al.*, 2006; Van Gerven *et al.*, 2005; Bertos *et al.*, 2004b), which involves less-reactive Ca-Mg silicates (Reactions (4.6)–(4.7)) (Huijgen *et al.*, 2005; Olajire, 2013, 2010).

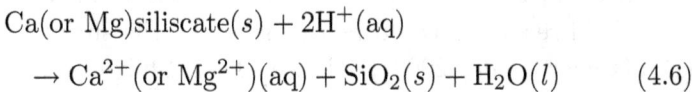

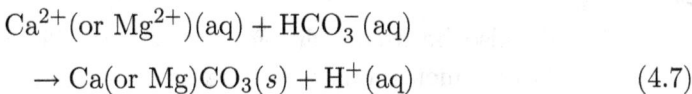

$$\text{Ca(or Mg)siliscate}(s) + 2H^+(aq)$$
$$\rightarrow Ca^{2+}(\text{or } Mg^{2+})(aq) + SiO_2(s) + H_2O(l) \tag{4.6}$$

$$Ca^{2+}(\text{or } Mg^{2+})(aq) + HCO_3^-(aq)$$
$$\rightarrow \text{Ca(or Mg)}CO_3(s) + H^+(aq) \tag{4.7}$$

The proposed mechanism of MC in Figure 4.2 indicates that carbonation reaction could occur in four routes: (1) $CaCO_3$ conversion

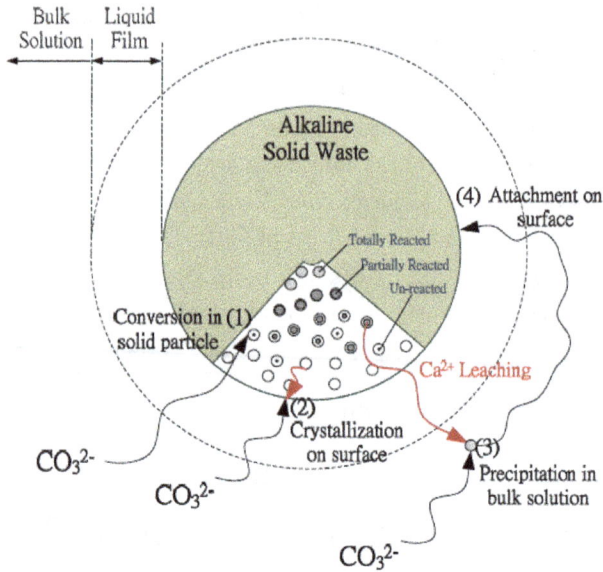

Figure 4.2 Proposed mechanism of MC reaction of alkaline solid wastes. Adapted from Pan *et al.* (2012).

inside the particle, (2) $CaCO_3$ crystallisation on surface, (3) $CaCO_3$ precipitation in bulk solution and (4) attachment on solid solution (Pan *et al.*, 2012). Transportation-controlled mechanism (diffusion of CO_2 and Ca^{2+}–ions to/from reaction sites), boundary-layer effects (dissolution of $Ca(OH)_2$ at the particle surface caused by diffusion through the precipitate coatings) and pore blockage are the core mechanisms that affect carbonation rate and extent (Huntzinger *et al.*, 2009a; Huijgen *et al.*, 2005; Huijgen and Comans, 2006; Uibu and Kuusik, 2009). The rate-limiting mechanism of heterogeneous solid–fluid reactions could be described by the shrinking core model (Huntzinger *et al.*, 2009a; Ki Lee, 2004; Shih *et al.*, 1999; Lekakh *et al.*, 2008).

MC could also be used to rapidly remediate the contaminated solids that have cementitious properties. Stabilisation/solidification technique enables the capsulation of toxic waste matter into solid bulk and decreases its mobilisation. The proposed mechanism includes following steps: diffusion of $CO_2(g)$ in air and into the solid phase, solvation of $CO_2(g)$ to $CO_2(aq)$ and hydration of $CO_2(aq)$ to H_2CO_3, dissociation of H_2CO_3 to H^+, HCO_3^- and CO_3^{2-}, releasing Ca^{2+} and SiO_4^{2-} ions from cementitious phases (Ca_3SiO_5, Ca_2SiO_4), nucleation and precipitation of $CaCO_3$ and calcium-silicate-hydrate (CSH) gel, and finally secondary MC by converting CSH gel to SH gel and $CaCO_3$ (Reaction (4.8)) (Bertos *et al.*, 2004a; Pan *et al.*, 2012). Rapid hardening occurs as a result of the reactions.

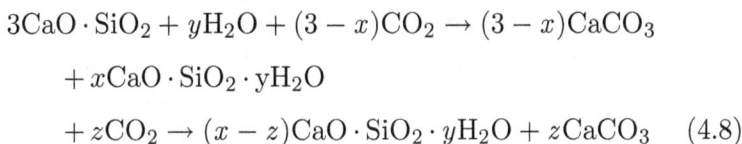

$$3CaO \cdot SiO_2 + yH_2O + (3 - x)CO_2 \rightarrow (3 - x)CaCO_3$$

$$+ xCaO \cdot SiO_2 \cdot yH_2O$$

$$+ zCO_2 \rightarrow (x - z)CaO \cdot SiO_2 \cdot yH_2O + zCaCO_3 \quad (4.8)$$

The extent and rate of carbonation depend mainly on the diffusive and reactive properties of CO_2 which in turn depend on the type of binder as well as its hydration degree, pore system and exposer conditions (Bertos *et al.*, 2004a).

4.3 Metallurgical Slag

Steelmaking industry generates substantial amounts of CO_2 (0.28–1 tCO_2/t-steel (Bonenfant *et al.*, 2008a), which globally accounts for 6–7% of CO_2 emissions (Doucet, 2009)) and slags (total production 315–420 Mt/yr (Sanna *et al.*, 2012a; Eloneva *et al.*, 2008b)). Integrated carbon steel production consists of ironmaking in the blast furnace (BF), steelmaking in the basic oxygen furnace (BOF) and continuous casting of steel billets, slabs and blooms (Bodor *et al.*, 2013). Steelmaking process produces mainly BOFS (also known as converter slag) (62%), electric arc furnace slag (EAFS) (29%), and ladle metallurgy slag (LS, also referred to as continuous casting slag) (9%) (Sanna *et al.*, 2012a; Gahan *et al.*, 2009). Blast furnace slag (BFS) is derived from iron production by melting the residues and coke ashes after grinding and separation of iron from ore (Sanna *et al.*, 2012a). Subsequent refining of stainless steel generates LS and argon oxygen decarburisation slag (AODS) as by-products. While these slags could be utilised in many industries, such as road construction, cement manufacturing or agriculture, it is not always economical or fails to fulfil the environmental regulations and thus end up being waste.

Metallurgical slags consist predominantly calcium silicates, but also other compounds and elements depending on the production process (Teir *et al.*, 2007a; Eloneva *et al.*, 2008b). Their annual total CO_2 uptake is approximately 171 Mt of CO_2 (Eloneva *et al.*, 2008a), around 0.6% of global CO_2 emissions from fuel combustion (Bobicki *et al.*, 2012). The main disadvantage to the MC is the milling requirement to generate sufficient reactive surface area, as these slags (except powdery AOD and Ladle slags) harden upon cooling in the form of monoliths (Bodor *et al.*, 2013). Research on utilisation of steelmaking slag for CO_2 sequestration has generally been focused on direct aqueous carbonation (DAC) and indirect aqueous phase carbonation (IAC).

Direct aqueous carbonation is generally carried out in a slurry phase (aqueous suspension contacted with gas phase) or via the thin film (humidified waste material in contact with gas phase) route (Baciocchi *et al.*, 2015). The E_{CO_2} values vary in very wide range (1.7–28.9%) and correspond to the slag type as well as operating conditions (Table 4.1). The carbonation kinetics and extent could be improved significantly via elevated pressures and temperatures as well as additives (Santos *et al.*, 2014) and UV treatment (Santos *et al.*, 2013a, Said *et al.*, 2015), but it also increases costs (up to €4000/t-slag) (Santos *et al.*, 2013a). Particle size is also an important variable, as most of the studies concentrate on using smaller fractions ($<150\,\mu$m) (Sanna *et al.*, 2012a; Huijgen *et al.*, 2005; Bonenfant *et al.*, 2008a; Lekakh *et al.*, 2008). According to estimations (Huijgen *et al.*, 2007), DAC of slag (200°C and 20 bar and 100% CO_2) costs €77/tCO_2 net avoided. Zingaretti *et al.* (2013) compared the material and energy requirements for slurry and thin film process routes. The total energy requirement for slurry-phase process, considering the operating conditions (T = 100°C; P = 10 bar; L/S = 5 w/w) and carbonation time of 1 h as well as grinding, milling and mixing of the slags with water, the heating and pumping of the slurry, the pressurisation of CO_2 and the solid/liquid separation of the slurry, would amount to 1350–2200 MJ/tCO_2 for different kinds of slags. As for the thin film route, the total energy requirements

Table 4.1 Main properties and carbonation conversions of iron and steelmaking slags (DAC — direct aqueous carbonation, IAC — indirect aqueous carbonation, RPB — rotating backed bed reactor, FB — fluidised bed reactor, T_{CO_2} — theoretical maximum CO_2 uptake, E_{CO_2} — experimental CO_2 uptake, η — carbonate conversion).

Slag Type	CaO, % MgO, %	$T_{CO_2}^1$, %	MC Route	CO_2, vol%	Operating Conditions	Performance, %	Reference
BFS	15–42 5–11	20–44	IAC pH swing	100	1. $T = 70°C$, CH_3COOH 2. $T = 30°C$, NaOH	$\eta = 74\%$ precipitation	(Eloneva et al., 2008a)
			DAC thin film	99.5	Batch: $P = 5$ bar; $t = 2$ h	$E_{CO_2} = 7\%$ $\eta = 15\%$	(Monkman and Shao, 2006)
BOFS	34–56 2–6	29–52	DAC slurry	100	Batch: $T = 100°C$; $P = 19$ bar; $t = 30$ min	$E_{CO_2} = 15.6\%$ $\eta = 74\%$	(Huijgen et al., 2005)
			DAC slurry	100	FB: $T = 60°C$; $P = 1$ bar; $t = 60$ min; $d < 44\,\mu m$	$E_{CO_2} = 29.0\%$ $\eta = 72.2\%$	(Chang et al., 2011)
			DAC slurry	100	RPB: $T = 65°C$; $P = 1$bar; $T = 30$ min; $d < 63\,\mu m$	$E_{CO_2} = 28.9\%$ $\eta = 93.5\%$	(Chang et al., 2012)
			DAC slurry	100	Column: $T = 25°C$; $P = 1$ bar; $t = 120$ min; $d < 44\,\mu m$	$E_{CO_2} = 28.3\%$ $\eta = 89.4\%$	(Chang et al., 2013b)

(Continued)

Table 4.1 (*Continued*)

Slag Type	CaO, % MgO, %	$T_{CO_2}^{1}$, %	MC Route	CO_2, vol%	Operating Conditions	Performance, %	Reference
			DAC slurry	98.9	RPB: T = 25°C; P = 1 bar; t = 1 min; d < 62 μm	E_{CO_2} = 10.3% η = 36.1%	(Pan et al., 2013)
			DAC slurry	30	RPB: T = 25°C; P = 1 bar; t = 20 min; d < 125 μm	E_{CO_2} = 27.7% η = 90.7%	(Chang et al., 2013a)
			DAC thin film	60	Batch: T = 650°C; P = 20 bar; t = 30 min; d < 80 μm	E_{CO_2} = 13.7% η = 30.2%	(Santos et al., 2012)
			DAC thin film	100	Batch: T = 50°C; P = 10 bar; t = 24 h; d < 150 μm	E_{CO_2} = 20.9%	(Baciocchi et al., 2015)
			DAC slurry	100	Batch: T = 100°C; P = 10 bar; t = 24 h; d < 150 μm	E_{CO_2} = 40.3%	(Baciocchi et al., 2015)
			IAC pH swing	13	1: T = 80°C, NH₄Cl; d < 2000 μm 2: T = 80°C	η = 48.1% extraction η = 70% precipitation	(Kodama et al., 2008)

Material			Technology		Conditions	Results	Reference
		24–48	IAC pH swing	100	1. T = 30°C; CH$_3$COOH; 2. T = 30°C; P = 1 bar; NaOH	η = 31–86% precipitation	(Eloneva et al., 2008b)
EAFS	25–47 4–19		DAC slurry	15	Batch: T = 20°C; P = 1 bar; t = 24 h; d = 38–106 μm	E$_{CO_2}$ = 1.7% η = 6.6%	(Bonenfant et al., 2008a)
			DAC slurry	15	Batch: T = 20°C; P = 1 bar; t = 65 min; d < 100 μm	E$_{CO_2}$ = 1.9–8.7%; η = 9.0–33.6%	(Uibu et al., 2011)
			DAC slurry	100	Batch: T = 25°C; P = 1 bar; t = 70 h; d = 150–250 μm	E$_{CO_2}$ = 3.0% η = 12%	(Lekakh et al., 2008)
			DAC slurry	100	Batch: T = 100°C; P = 10 bar; t = 24 h; d < 150 μm	E$_{CO_2}$ = 28.0% η = 72.3%	(Baciocchi et al., 2015)
			DAC thin film	100	Batch: T = 50°C; P = 10 bar; t = 24 h; d < 150 μm	E$_{CO_2}$ = 17.6% η = 45.4%	(Baciocchi et al., 2015)

(*Continued*)

Table 4.1   (*Continued*)

Slag Type	CaO, % MgO, %	$T_{CO_2}{}^1$, %	MC Route	CO_2, vol%	Operating Conditions	Performance, %	Reference
LS	42–58 6–15	42	DAC slurry	15	$T = 20°C$; $P = 1$ bar; $d = 38$–$106\ \mu m$	$E_{CO_2} = 24.7\%$ $\eta = 54.1\%$	(Bonenfant et al., 2008a)
			DAC slurry	15	Batch: $T = 20°C$; $P = 1$ bar; $t = 65$ min; $d < 100\ \mu m$	$E_{CO_2} = 4.6\%$ $\eta = 13.9\%$	(Uibu et al., 2011)
AODS	41–61 4–8	31–54	DAC slurry	99.5	$T = 50°C$; $P = 1$ bar; US = 24 kHz; $t = 240$ min; $d = 63$–$200\ \mu m$	$\eta = 49\%$	(Santos et al., 2013a)
			DAC slurry	99.5	$T = 50°C$; $P = 1$ bar; $t = 240$ min; $d = 63$–$200\ \mu m$	$\eta = 30\%$	(Santos et al., 2013a)
			DAC slurry	99.5	$T = 50°C$; US = 24 kHz; $MgCl_2$; $t = 240$ min	$E_{CO_2} = 27\%$	(Santos et al., 2014)

DAC slurry	99.5	T = 50°C; US = 24 kHz; t = 240 min	E_{CO_2} = 19%	(Santos et al., 2014)
DAC slurry	99.5	T = 90°C; P = 6 bar; t = 120 min; d < 500 μm	η = 40.6%	(Santos et al., 2013c)
DAC thin film	20	T = 30°C; P = 1 bar; t = 144 h; d < 500 μm	η = 24.2%	(Santos et al., 2013c)

Remarks + high CO_2 sequestration capacity; generated in large quantities; generated near CO_2 source; carbonation improves mechanical and environmental parameters of slag; possible applications: PCC.

− must undergo milling (except AODS and LFS); high T&P, US (€4,000/t slag (Santos et al., 2013a)), or additives are needed for acceptable conversion.

Cost: €77/tCO_2 net avoided (DAC) (Huijgen et al., 2007); IAC with PCC production: 300 kWh/tCO_2 (Kodama et al., 2008) or €1990/t$CaCO_3$ for chemicals only (Eloneva, 2010).

[1]T_{CO_2} — theoretical maximum CO_2 uptake expressed in wt.% was calculated using a modified Steinour formula:
$T_{CO_2} = 0.785 \times (\%CaO - 0.53 \times \%CaCO_3 - 0.7 \times \%SO_3) + 1.091 \times \%MgO + 0.71 \times \%Na_2O + 0.468 \times (\%K_2O - 0.632 \times \%KCl)$.

calculated considering the operating conditions (T = 50°C, P = 10 bar; L/S = 0.3 w/w), a carbonation time of 1 h, would amount to 2100–2570 MJ/tCO_2. In this case, the total requirements include the grinding and milling of the slags, the pressurisation of CO_2 and the heating and operation of the rotary drum carbonation reactor (Baciocchi *et al.*, 2015).

The utilisation of wastes for CO_2 sequestration would be more appealing to the industry if the benefits of MC and waste valorisation were combined symbiotically. The IAC of steelmaking slags is especially interesting due to the production of high-purity products such as high-purity precipitated $CaCO_3$ (PCC) (Kodama *et al.*, 2008; Eloneva *et al.*, 2008b; Teir *et al.*, 2007a; Chiang *et al.*, 2014). Several extraction agents including HNO_3, H_2SO_4, NaOH (Doucet, 2009), NH_4Cl (Kodama *et al.*, 2008; Eloneva *et al.*, 2012; Said *et al.*, 2013; Mattila and Zevenhoven, 2014), NH_4NO_3 (Said *et al.*, 2013), CH_3COONH_4 (Said *et al.*, 2013) and CH_3COOH (Eloneva *et al.*, 2008b; Teir *et al.*, 2007a; Chiang *et al.*, 2014), NH_4NO_3 (Eloneva *et al.*, 2012) NH_4HSO_4 (Dri *et al.*, 2013) have been considered for the IAC route.

Slag2PCC (Figure 4.3a), a MC concept, that uses steel converter slag as a raw material to produce PCC, has been developed on the level of theoretical process modelling and batch experiments (Said *et al.*, 2013; Eloneva *et al.*, 2008b) and a small-scale demonstration plant is opened at Aalto University (Mattila and Zevenhoven, 2014). The slag2PCC process is based on the re-circulation of an aqueous ammonium salt solution between two stages: the extraction of Ca from slag and the carbonation of the dissolved Ca with gaseous CO_2. The concept focuses on the direct use of purified flue gases as well as the possibility to operate the system at ambient temperature and pressure. Based on the global amount of BOFS produced, slag2PCC process would be able to capture 19.1 Mt CO_2 annually (0.1 kg CO_2 per kg slag) (Mattila and Zevenhoven, 2014). Comparing this approach with the conventional PCC production method (SpecialChem, 2013) by means of a cradle-to-gate life cycle assessment (LCA), indicated that although the Slag2PCC process has negative CO_2 emissions and is environmentally more benign, the

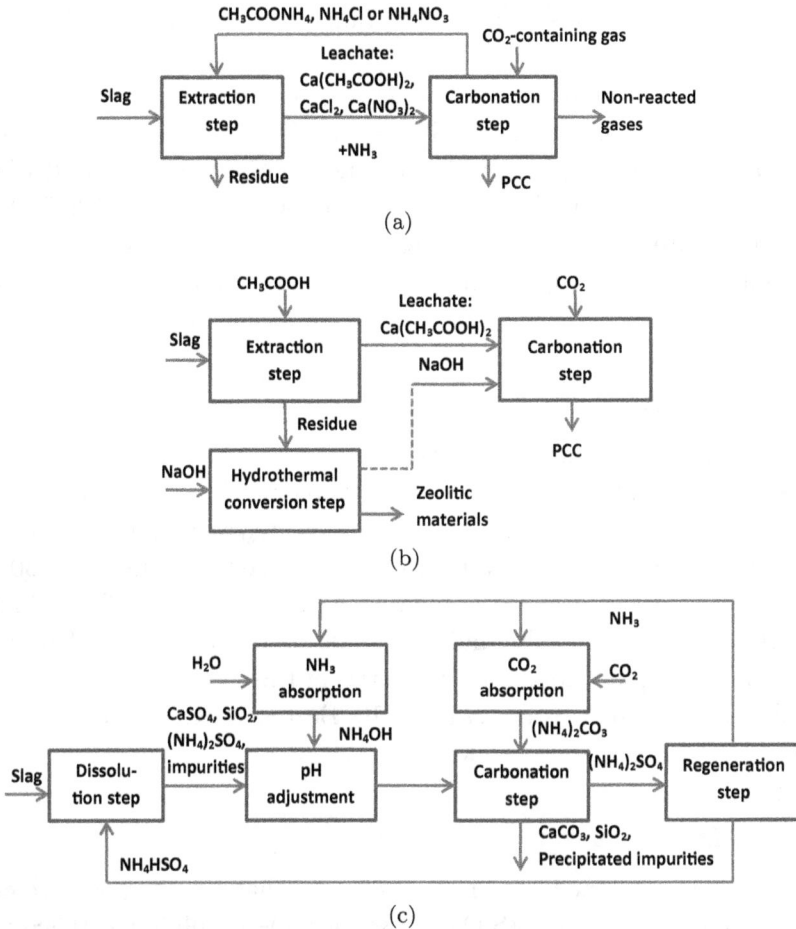

Figure 4.3 MC concepts for slag valorisation. (a) Slag2PCC concept (modified from Mattila and Zevenhoven, 2014). (b) Two-way valorisation of BFS (modified from Chiang *et al.*, 2014). (c) Multi-step closed loop MC process (modified from Dri *et al.*, 2014).

purification of this PCC would pollute more than the traditional method (Mattila *et al.*, 2014).

A three-stage process has been developed for two-way valorisation of metallurgical slag. According to the concept (Figure 4.3b), Ca is selectively extracted by the means of an organic acid in the first stage, following by carbonation of the Ca-solution to produce PCC in the second stage and to form zeolites by hydrothermal conversion of the extracted solids in alkali solution in the third stage (Chiang et al., 2014).

A closed loop multi-step process (Figure 4.3c) was developed to extract Ca^{2+} from steel slag with NH_4HSO_4 solution to form solid $CaSO_4$, which after pH adjustment and precipitation of impurities was contacted with $(NH_4)_2CO_3$ (from CO_2 capture with NH_3) to precipitate $CaCO_3$ (Dri et al., 2013, 2014). The carbonation efficiencies achieved for different slags ranged from 59 to 74% (Dri et al., 2014).

4.4 MSWI Ashes

MSWI is a waste management technology that offers the reduction of mass (up to 70%), volume (up to 90%) and organic content of wastes, as well as disinfection and potential energy recovery, termed waste-to-energy (WtE). The MSWI also generates solid residues (20–30% of the original waste mass), including APC residues and BA, which are the final sinks of salts and numerous toxic and regulated heavy metals. Thus, the valorisation potential of these materials is limited (Bobicki et al., 2012; Wang et al., 2010; Bertos et al., 2004b; Li et al., 2007; Costa et al., 2007; Bodor et al., 2013).

4.4.1 *MSWI BA*

MSWI BA, classified generally as a non-hazardous waste, consists mainly of silicates (SiO_2, $Ca_2Al[AlSiO_7]$), sulphates ($CaSO_4$, $CaSO_4 \cdot 2H_2O$, $Ca_6Al_2(SO_4)_3(OH)_{12} \cdot 26H_2O$), carbonates ($CaCO_3$), but also metal oxides, hydroxides ($Ca(OH)_2$, Fe_2O_3, Fe_3O_4) and chlorides (Costa et al., 2007; Sanna et al., 2012a; Bobicki et al., 2012). In lack of suitable utilisation options, the usual practice to date has been natural weathering (NW), with the aim of decreasing leaching to environmentally tolerable levels before landfilling (Bodor et al., 2013). Too low contents of Ca and Mg does not allow substantial CO_2 sequestration, but the chemical stability and leaching behaviour could be improved by MC technique (Bertos et al., 2004b; Costa et al., 2007) for further valorisation as a secondary building material (Bobicki et al., 2012; Rendek et al., 2006).

The accelerated MC of MSWI BA has been studied to track the mineralogical changes similar to NW and to reduce alkalinity and trace metal mobility (Van Gerven *et al.*, 2005; Meima *et al.*, 2002; Polettini and Pomi, 2004; Bertos *et al.*, 2004b). The common process route has been thin film carbonation (L/S = 0.2–0.3 w/w) in a CO_2 rich environment (10–100% vol CO_2) at moderate temperatures (30–50°C) for longer time period (up to 7 days) in static conditions (e.g. thinly spread layers) (Rendek *et al.*, 2006; Bertos *et al.*, 2004b). The general precept is to maintain the optimal temperature for CO_2 dissolution as well as for driving carbonation reaction kinetics. The moisture content in the mixture should be sufficient for aqueous carbonation reaction, but thin enough to limit the diffusion distance between carbonate ion and the reaction zone (Bodor *et al.*, 2013). The MSWI BA has generally shown CO_2 uptake on the order of 3.0–6.5 wt.% (Bertos *et al.*, 2004b; Gunning *et al.*, 2010; Costa *et al.*, 2007; Rendek *et al.*, 2006) (Table 4.2). Santos *et al.* (2013b) concluded from comparative study of heat treatment, ageing and accelerated MC that the latter is the most effective stabilisation treatment for MSWI BA in a view of heavy metal/metalloid leaching. MC has shown improvement in the leaching behaviour of certain metals, such as Cu, Pb and Zn, but detrimental effect on the leaching of Cr, Mo and Sb (Arickx *et al.*, 2006; Van Gerven *et al.*, 2005; Arickx *et al.*, 2010; Cornelis *et al.*, 2006; Baciocchi *et al.*, 2010; Rendek *et al.*, 2006). The practical set-up for accelerated carbonation can be achieved by directing the stack gas in counter-current over a MSWI BA layer on a moving belt.

4.4.2 *APC Residues*

Limiting emissions into the atmosphere has shifted the contaminants from gaseous emissions to the solid APC residues. Solid residues generated from flue gas treatment contain a mixture of fly ash and excess sorbent material (activated carbon and lime). APC residues are categorised as hazardous wastes due to the elevated content of lime (pH > 12), heavy metals (Zn, Pb, Cd, Cr, Cu, Hg, Ni), chlorinated compounds and soluble salts (Bertos *et al.*, 2004b;

Table 4.2 Main properties and carbonation conversions of MSWI ashes.

Ash Type	CaO, % MgO, %	T_{CO_2}, %	MC Route	CO_2, vol%	Operating Conditions	E_{CO_2} %	Reference
MSWI BA	22–53 2.8	25	DAC thin film	100	Ambient T; P = 3 bar; RH = 65%; t = 2.5 h; d < 710 μm	3.2	(Bertos et al., 2004b)
			DAC thin film	100	ambient T; P = 2 bar; t = 72 h	4.0	(Gunning et al., 2010)
APC residue	36–60 1–2.5	50–58	DAC thin film	100	ambient T; P = 3 bar; RH = 65%; t = 2.5 h; d < 212 μm	7.3	(Bertos et al., 2004b)
			DAC thin film	20	T = 20–30°C; t = 50–150 min; d_{mean} = 66 μm	8–12	(Sun et al., 2008)
			DAC thin film	100	T = 30–50°C; P = 1–10 bar		(Baciocchi et al., 2009b)
			DAC thin film	100	Ambient T; P = 2 bar; t = 72 h		(Gunning et al., 2010)
Remarks	+produced in large quantities (MSWI BA); produced near CO_2 source; carbonation reduces pH and leaching of hazardous elements for safer landfill; Grinding not required. −low carbon sequestration capacity (MSWI BA).						

Bobicki *et al.*, 2012; Sanna *et al.*, 2012a). MC of lime-rich APC residues as a potential CO_2 sequestration option could reduce the pH levels to regulatory limits (pH < 9.5) (Baciocchi *et al.*, 2006a, 2006b; Bertos *et al.*, 2004a, 2004b).

The main reactive species of APC residues include $Ca(OH)_2$ and CaClOH, which together with high specific surface areas attribute to the fast CO_2 uptake kinetics (Bertos *et al.*, 2004b). Leaching decreases mainly due to the precipitation of carbonates which leads to the decrease of porosity and the lowering of pH. The average E_{CO_2} uptake is in the range of 7–25 wt.% (0.07–0.25 tCO_2/t-APC) (Baciocchi *et al.*, 2009b; Bertos *et al.*, 2004b; Sanna *et al.*, 2012a; Baciocchi *et al.*, 2006b; Sun *et al.*, 2008) (Table 4.2). According to Baciocchi *et al.* (2009b) dry route (1 bar, 10–50% CO_2, 350–500°C) results faster reaction kinetics as compared to thin film (1–10 bar, 100 vol% CO_2, 30–50°C, L/S > 0.6) at similar maximum carbonation conversion (65%).

4.5 Power Plant Ashes

Globally, coal-fired power plants are still the largest industrial source of CO_2 (14,800 $MtCO_2$, in 2013) which also generate a large quantity of fly ash (FA, 600 Mt in 2009) (Bobicki *et al.*, 2012; Sanna *et al.*, 2012a; IEA 2015). Coal FA is typically a fine powder (d = 10–15 μm), which consists of amorphous aluminosilicate glass matrix ($Si_xAl_yO_z$) and recrystallised silica minerals (Sanna *et al.*, 2012a). The amount of FA put to beneficial use (as a cement additive) varies by country, the potential of FA for MC worldwide is estimated to be ~7 Mt carbon annually (Reynolds *et al.*, 2014). The bituminous coal FA contains 1–3% of MgO and 5–10% CaO and its maximum CO_2 sequestration potential is relatively low ~9 wt.% (Table 4.3). Alkaline components generated in the flue gas desulfurisation process enhance its potential to be used as a Ca source for MC (Han *et al.*, 2015). Also, low-grade fossil fuels such as lignite coal and oil shale (OS) generate Ca-rich FA (12–30% free lime, depending on combustion regimes PF or CFB) that could be attractive candidates for MC (Uibu *et al.*, 2010; Bauer *et al.*, 2011; Back *et al.*, 2008). Using wood in coal power plants

Table 4.3 Main properties and carbonation conversions of power plant ashes.

Ash Type	CaO, % MgO, %	T$_{CO_2}$, %	MC Route	CO$_2$, vol%	Operating Conditions	E$_{CO_2}$ %	Reference
Coal FA	1.3–10 1–3	6–9	DAC slurry	100	T = 20–60°C; P = 10–40 bar; t = 18 h; d$_{mean}$ = 40 μm	2.6	(Montes-Hernandez et al., 2009)
			DAC slurry	100	T = 30°C, 90°C; P = 10–40 bar; t = 18 h; d$_{ground}$ <150 μm	3.6-7.2	(Nyambura et al., 2011)
			DAC slurry		T = 90°C; P = 40 bar; bulk ash; t = 2 h	6.5	(Muriithi et al., 2013)
			NW	0.0365	ambient T&P; wet deposited ash; t = 20 y	6.8	(Muriithi et al., 2013)
Lignite FA	27.5 6.5	43	DAC slurry	10	T = 75°C, P = 1 bar; t = 4.5 h; d < 250 μm	23.0	(Back et al., 2008)
			DAC slurry	100	T = 30–80°C; NaCl (1–25 g/L); pH = 5–9; d$_{ground}$ = 30–125 μm; t = 10–50 min	7.1	(Mayoral et al., 2013)

OS FA	38–50 5–12	26–49	DAC thin film	10–20	Ambient T&P; t = 120 min	21.1	(Bauer et al., 2011)
			DAC slurry	15	ambient T&P; t = 65 min	29	(Uibu et al., 2011)
			NW	0.0365	ambient T&P; wet deposited ash; t = 8 w	2.2	(Uibu et al., 2009)
WA	24–49 8–9	50	DAC thin film	100	Ambient T, P = 2 bar; H_2O to form paste; t = 72 h	8	(Gunning et al., 2010)

Remarks +produced in large quantities (coal FA); produced near CO_2 source; grinding usually not required; high CO_2 sequestration capacity (OS FA); −low CO_2 sequestration capacity (Coal FA); Waste available in few areas (OS FA);

Cost: \$11–21/tCO_2 at mineralisation capacity of 0.1–0.2 tCO_2/t-FA (Reddy et al., 2010, 2011).

and combined heat and power plants generates wood ash (WA). WA contains 45% of CaO (Gunning *et al.*, 2010).

The MC is mainly carried out under mild process conditions using either water (Muriithi *et al.*, 2013; Montes-Hernandez *et al.*, 2009; Uibu *et al.*, 2010; Back *et al.*, 2008) or brine (Soong *et al.*, 2005; Nyambura *et al.*, 2011) as the reaction medium. The NW over longer time period has also been tested (Muriithi *et al.*, 2013; Uibu *et al.*, 2009). The experimental CO_2 uptakes depend on the ash type and range from 0.026 (Montes-Hernandez *et al.*, 2009) to 0.072 t/t (Nyambura *et al.*, 2011) for coal FA and from 0.055–0.230 t/t for lignite FA (Uliasz-Bocheńczyk *et al.*, 2009; Back *et al.*, 2008; Mayoral *et al.*, 2013; Wee, 2013) up to 0.29 t/t for OS FA (Uibu *et al.*, 2011).

However, as the FA with an average CO_2 uptake of 5% could sequester only 0.25% of CO_2 emissions from coal-fired power plants (Sanna *et al.*, 2012a; Bobicki *et al.*, 2012; Montes-Hernandez *et al.*, 2009), the focus has shifted to lowering the alkalinity of leachates (Uibu *et al.*, 2009) and the stabilisation of the contaminants such as Cd, Pb, Cr, As, Se, Al and S, which are generally co-precipitated with carbonated FA (Wee, 2013). A demonstration scale MC process was developed and tested by reacting FA with flue gases in a FB reactor (Figure 4.4) at Jim Bridger Power Plant, a 2,120 MW coal-fired power plant, located near Point of Rocks, USA. The MC results indicated that multiple pollutants can be removed from flue gas (CO_2, SO_2, Hg) and the soluble trace elements in FA are sifted to less leachable mineral fractions (Reynolds *et al.*, 2014).

For further valorisation, OS FA could be utilised as a low-cost Ca-source for PCC production in IAC process (Figure 4.5) (Velts *et al.*, 2011). The lime depleted residue is able to bind an additional amount of CO_2 on account of Ca-silicates (Uibu *et al.*, 2011). High brightness PCC of 96% purity was achieved in ambient conditions and using water as a Ca leaching agent.

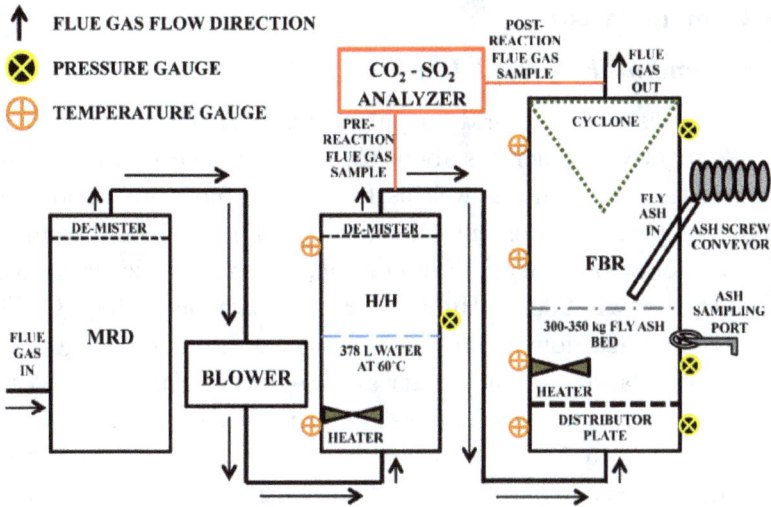

Figure 4.4 Schematic of MC process design for coal FA (adapted from Reynolds *et al.*, 2014).

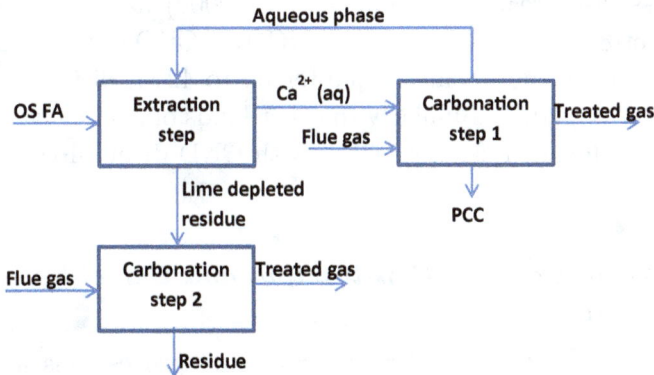

Figure 4.5 MC concept for OS FA valorisation (modified from Uibu *et al.*, 2011).

4.6 Cement Wastes

4.6.1 *Cement Kiln and Bypass Dust*

Cement industry (the world output of 2.8 Gt cement is expected to increase to 4.0 Gt) generates about 5% of global CO_2 emissions (from calcining the limestone as well as from fossil fuel combustion) and millions of tonnes of cement kiln dust (CKD) (0.15–0.20 tonnes of CKD per tonne of cement) (Sanna *et al.*, 2012a; Bobicki *et al.*, 2012; Huntzinger *et al.*, 2009a; Huntzinger and Eatmon, 2009). CKD is considered a hazardous waste due to its causticity and it is generally disposed into landfills (Huntzinger and Eatmon, 2009; Bobicki *et al.*, 2012). As CKD is separated before kiln firing, it contains considerable amount (about 46–57%) of uncalcined $CaCO_3$ (Sanna *et al.*, 2012a; Huntzinger *et al.*, 2009a; Bobicki *et al.*, 2012). Cement bypass dusts (CBD) on the other hand has a much lower carbonate content and thus higher CO_2 binding capacity (0.5 tCO_2/t-CBD) because it is removed after kiln firing (Gunning *et al.*, 2009). Slurry-phase MC at ambient conditions (Huntzinger *et al.*, 2009a) and thin film route in a pressurised reactor (Gunning *et al.*, 2009) have resulted E_{CO_2} uptakes of 8–25 wt.% (i.e. 0.08–0.25 tCO_2/t-CKD,CBD), indicating that CKD could potentially sequester up to 42 Mt of CO_2 annually (Bobicki *et al.*, 2012). Similarly to other kinds of alkaline waste, MC reduces the health risks associated with CKD disposal (Huntzinger *et al.*, 2009b).

4.6.2 *Waste Cement, Construction and Demolition Wastes*

The waste cement from aggregate recycling process has also been considered for MC due to the CO_2 sequestration potential (61 Mt CO_2 considering the annual waste concrete production of 1,100 Mt (Bobicki *et al.*, 2012)) although most of it is currently utilised in construction industry (Sanna *et al.*, 2012a). These utilisation options could be combined by using CO_2-activated hardening process to produce building materials (with maximum E_{CO_2} 16.5%, at 100% CO_2 and 4 bars (Teramura *et al.*, 2000), Table 4.4). The studies have indicated that the energy required for concrete curing will

Table 4.4 Main properties and carbonation conversions of cement wastes.

Waste Type	CaO, % MgO, %	T_{CO_2}, %	MC Route	CO_2, Vol%	Operating Conditions	E_{CO_2} %	Reference
CKD	34–48 1–1.5	10–30	DAC thin film	5–15	Ambient T&P; t = 8 h	8–18	(Huntzinger et al., 2009a)
			DAC thin film	100	Ambient T, P = 2 bar, H_2O to form paste; t = 72 h	10	(Gunning et al., 2010)
CBD	66 1	50	DAC thin film	100	Ambient T, P = 2 bar, H_2O to form paste; t = 72 h	25	(Gunning et al., 2010)
Waste cement	25–63 0.3–2	20	DAC thin film	0.03–100	T = 20°C; P = 1–4 bar; t = 0.8–10 h; d < 1.8 mm;	1.6–16.5	(Teramura et al., 2000)
			DAC thin film	20	T = 20°C; t = 60 min	8.9	(Kashef-Haghighi and Ghoshal 2009)

Remarks: +generated in large quantities; generated near CO_2 source (CKD, CBD); carbonated product can be re-used in cement manufacturing, aggregates to produce PCC etc.; CKD, CBD has fine particle size;
−Particles size reduction required for cement waste; low carbon sequestration capacities (CKD).

be significantly reduced by switching from conventional steam and autoclave curing to accelerated CO_2 curing with low pressure flue gas. Furthermore, the resulting concrete products (concrete blocks, bricks, modular building elements) are characterised by enhanced resistance to surface permeation, freeze-thaw cycling and sulphate attack (Kashef-Haghighi *et al.*, 2015). According to Kashef-Haghighi and Ghoshal (2013), about 1.5 Mt of CO_2 could be sequestered on the basis of an average carbonation efficiency of 18% and the estimated 16.4 Mt of cement used in concrete products in the US in 2004. The same process could be applied to other kinds of cementitious wastes such as EAF slag, etc. (Monkman and Shao, 2006).

The IAC of waste cement powder has also been studied. Katsuyama *et al.* (2005) and Iizuka *et al.* (2004) extracted Ca^{2+} from cement waste by pressurised CO_2 (30 bar) and used sequent carbonation at reduced pressures (1 bar) to produce high-purity $CaCO_3$ (the estimated costs per 1 metric tonne of $CaCO_3$ were US\$136–323 depending on purity (Katsuyama *et al.*, 2005)). The electrochemical method has also been studied as a potential way to recover the chemicals used in accelerating the reaction. Shuto *et al.* (2014) proposed a new MC process with waste cement powder via regeneration of alkali and acid by electrodialysis (Figure 4.6). For potassium acetate, the power consumption is 400 kJ/mol-CO_2 and the cost for MC is about US\$175 per metric tonne of CO_2 (Shuto *et al.*, 2015).

4.7 Mining Tailings, Asbestos-Containing Materials

The utilisation of Mg-rich tailings from mineral processing of different types of ores (copper–nickel–platinum group elements, asbestos, podiform chromite, diamondiferous kimberlite) in MC is one of the few realistic choices for large-scale above-ground CO_2 sequestration. These tailings residues, besides being abundant and already mined and milled, possess high porosity (Bobicki *et al.*, 2012; Wilson *et al.*, 2009; Bodor *et al.*, 2013). In addition, the hazardous asbestos tailings could be remediated by MC treatment (Bobicki *et al.*, 2012).

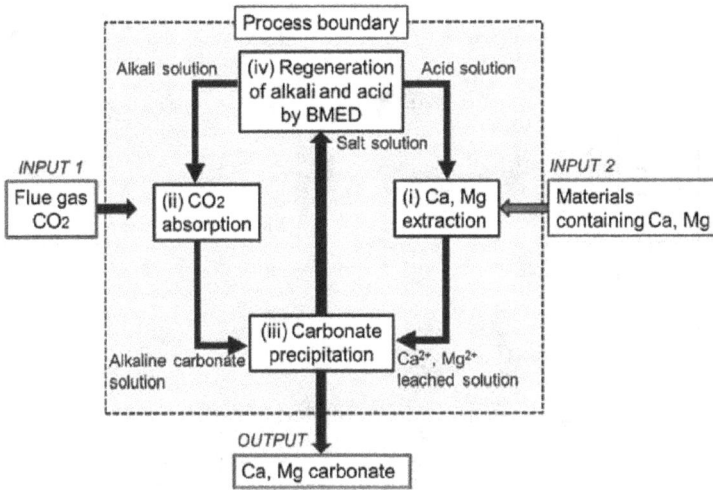

Figure 4.6 Flow diagram of a MC process regeneration of alkali and acid by electrodialysis, adapted from Shuto *et al.* (2014).

The ultramafic wastes contain Mg-rich minerals such as olivine $((Mg,Fe)_2SiO_4)$ and serpentine $((Mg,Fe)_3Si_2O_5(OH)_4)$. The tailings from chrysotile $(Mg_3(Si_2O_5)OH_4)$ processing are classified as hazardous wastes due to residual asbestos. The asbestiform nature of the mineral can be destroyed by MC. Thus, this approach has a potential to be used also for remediating asbestos wastes from the construction and demolition industry (cement–asbestos) instead of costly thermal treatments (Bodor *et al.*, 2013). Wilson *et al.* (2009) and Oskierski *et al.* (2013) have studied the NW of old waste piles (Table 4.5). According to the estimations of Wilson *et al.* (2009) the chrysotile in the tailings piles had carbonated $\sim$0.3% per year. The effect of potential field conditions (pore saturation, watering regime, temperature, CO_2 diffusion and dissolved oxygen) and a selection of natural and chemical enhancers (sulphide minerals, brucite, chelate ligands, ionic liquids and carbonic anhydrase enzyme) on the carbonation of chrysotile and nickel mining was studied by Assima *et al.* (2014d). Watering of the residues was found critical for mediating magnesium leaching and CO_2 absorption. Increasing temperature enhanced CO_2 uptake whereas presence of oxygen caused undesirable passivation

Table 4.5 Main properties and carbonation conversions of mine tailings and RM.

Waste Type	CaO, % MgO, %	T_{CO_2}, %	MC Route	CO_2, Vol%	Operating Conditions	E_{CO_2} %	Reference
Asbestos tailings	0.2–12 16–39	29–43	DAC thin film	9.96	Ambient T&P; half-saturated with H_2O; t = 17 h	0.31–0.37	(Assima et al., 2014a)
			DAC	100	T = 290°C, 5 bar, t = 5 h	26	(Ryu et al., 2011b)
Ni tailings	0.1–3.4 14–40	27–43	DAC thin film	9.96	Ambient T&P; half-saturated with H_2O; t = 17 h	0.13–1.12	(Assima et al., 2014a)
			IAC	100	(1) Step (T = 70°C; L(4M HCl, HNO_3)/S = 10; ground d < 0.5 mm; t = 2 h) (2) Step (T = 30°C; NaOH; t = 0.5 h)	29	(Teir et al., 2007b)

RM	1–47	<1	7–19	DAC thin film	100	T = 20°C; P = 3.5 bar; ground d_{mean} = 30 μm; t = 12 h	5.3	(Yadav et al., 2010)
				DAC	100	Ambient T&P; ground d = 0.1–160 μm; multi-cycle (each 5 h)	7.2	(Sahu et al., 2010)
				DAC slurry	15	Ambient T&P; d < 1000 μm; t = 24 h	4.15	(Bonenfant et al., 2008b)

Remarks

+carbonation destroys the asbestos nature (asbestos tailings, Ni tailings in chrysotile present); grinding not required (asbestos and Ni tailings); large quantities produced in localised areas; carbonation stabilises RM disposal

−low carbon sequestration E_{CO_2}; too expensive to achieve high carbonate conversion (Asbestos and Ni tailings); bicarbonates generated instead of carbonates (RM).

Cost: $147/tCO$_2$ for DAC of RM (Sahu et al., 2010).

by iron (III) hydroxides, which could be reduced through addition of dilute CDTA chelator (Assima *et al.*, 2014c, 2014d). A positive correlation between CO_2 breakthrough time and liquid saturation was exposed by dynamic tests with Ni mining residue. The CO_2 uptake was substantially stimulated under partially saturated conditions (Assima *et al.*, 2014a, 2014b).

According to Larachi *et al.* (2010) low pressure direct gas–solid carbonation of chrysotile via surface impregnation of super-basic sites or amorphisation/dehydroxylation leads to rather poor conversions (3.3% after 10 h at 375°C). Partial dehydoxylation and steam mediation could enhance the conversions (uptakes $> 0.7\,CO_2$ moles per Mg mole at 130°C) (Larachi *et al.*, 2012). Ryu *et al.* studied direct aqueous carbonation of chrysotile under subcritical conditions in alkali solution (Ryu *et al.*, 2011a) and tremolite ($Ca_2Mg_5Si_8O_{22}(OH)_2$) at 290°C and 5 bar CO_2 (Ryu *et al.*, 2011b) achieving substantial carbonation conversions (up to 60 wt.% of the final material). However, it is obvious that CO_2 sequestration by Mg-rich tailings requires elevated T&P conditions or pre-treatment, similar to serpentine treatment, in order to achieve substantial carbonate conversions (Table 4.5).

Integrating Ni mining operations with CO_2 sequestration and waste valorisation could in principle improve the feasibility of marginal Ni projects (Hitch *et al.*, 2009), but needs further development for cost reduction. This can be achieved by producing hydromagnesite of 93–99% purity (Teir *et al.*, 2007b), which would require US$600–1,600/t$CO_2$ for chemicals (Teir *et al.*, 2009) (Table 4.5) and a carbon regulatory framework, counting a cap-and-trade scheme with a high enough carbon price (Bobicki *et al.*, 2012).

4.8 RM

RM is a caustic by-product of alumina extraction from bauxite ore via Bayer process (1.0–1.5 tonnes RM per 1 tonne of alumina, 70 Mt annually (Dilmore *et al.*, 2007; Yadav *et al.*, 2010)), categorised as hazardous waste due to its high alkalinity. The RM is a mixture of liquid in chemical equilibrium with fine solids (20–80%) and its average chemical composition is in the range of Fe_2O_3 (7–72%), Al_2O_3 (2–33%), SiO_2 (1–50%), Na_2O (1–13%), CaO (1–47%) and TiO_2 (3–23%) (Sahu et al., 2010; Bodor et al., 2013). The pH averages at 11.3 ± 1.0, which is mainly contributed by the alkaline solids (hydroxides, carbonates, aluminates) and the residual NaOH in the liquid phase.

RM is an abundant hazardous material that could be re-mediated by MC in addition to CO_2 sequestration (Dilmore et al., 2007; Yadav et al., 2010). The carbonation products are applicable in brick, cement, fertilisers and plastics industry as well as wastewater treatment (Bonenfant et al., 2008b). MC of RM is principally carried out in slurry phase at ambient temperatures and pressures (Yadav et al., 2010; Sahu et al., 2010; Bonenfant et al., 2008b) to achieve sequestration capacity of 0.04–0.05 tCO_2/t RM (Table 4.5). The CO_2 sequestration cost is at US\$147/tCO_2 by rough estimation (Sahu et al., 2010). At Kwinana in Western Australia, MC has been employed as a pre-deposition treatment method. Gaseous CO_2 from nearby ammonia plant is mixed with RM in pressurised vessels, to reduce pH of the slurry and capture 0.030–0.035 tCO_2 per tonne of RM (GCCSI, 2011).

As the key alkalinity source in RM is NaOH in the liquid phase, the main carbonation products are soluble Na-carbonates which provide a less stable CO_2 sequestration option as compared to Ca–Mg carbonates (Sahu et al., 2010). Mixing RM with brine solution (saline wastewater rich in Ca^{2+} and Mg^{2+}) before carbonation is a more permanent option (Johnston et al., 2010; Dilmore et al., 2007). According to Si et al. (2013) an estimated 100 Mt of CO_2 have been sequestered through the NW of historically deposited RM (6 Mt annually). By utilising appropriate technologies for incorporating binding cations into RM, approximately 6 Mt of additional CO_2 could be sequestered while the RM is simultaneously remediated (Si et al., 2013).

4.9 Alkaline Paper Mill Wastes

The alkaline paper mill wastes from pulp and paper industry contain up to 82% of free CaO. The MC of paper mill wastes is dually beneficial as the CO_2 emissions from paper mills could be reduced and the precipitated $CaCO_3$ could be utilised as a value-adding product in the pulp and paper industry. (Perez-Lopez et al., 2008; Gunning et al., 2009; Bobicki et al., 2012). Direct aqueous carbonation of alkaline paper mill wastes has resulted E_{CO_2} of 16–27% (i.e. 0.16–0.27 tCO_2/t) at 20–30°C and 2–10 bar over 2–72 h (Perez-Lopez et al., 2008; Gunning et al., 2010).

4.10 MC Cost Assessment and Demonstration Projects

A number of large-scale industrial wastes have been considered as sorbents for MC, but regardless of some benefits (avoiding costs for mining and transport) in the current stage, they cannot compete with geological storage on the basis of CO_2 sequestration capacity and costs (GCCSI, 2009). The ones that are available in large quantities tend to have low CO_2 binding capacity (coal FA, MSWI BA) or require expensive processing (grinding, additives, elevated T&P) conditions (metallurgical slags, mining tailings). The ones that appear to be more suitable as CO_2 sorbents (APC residues, APMW, OS FA) are only available locally or in small scale. However, the countries that lack geological storage (Finland, Estonia), should consider these options as well (for instance OS FA could capture 10–12% of CO_2 emitted from OS power plants (Uibu et al., 2009)).

The costs of DAC estimations roughly based on laboratory experiments are in the range of US$8–104 per tCO_2, depending on the type of waste and operation conditions (Stolaroff et al., 2005; Huijgen et al., 2007). Using the IAC route with production of commercial by-products including PCC (from steel slags (Mattila and Zevenhoven 2014; Mattila et al., 2014), OS FA (Velts et al., 2011, 2014)) or hydromagnesite (from Mg-rich mining tailings (Teir et al., 2009)) would require additives (HCl, HNO_3, CH_3COOH, NaOH) for a price of $600–4,500 per tCO_2 without regeneration. The additives regeneration would emit 2.6–3.5 times more CO_2 than bound in the MC process (Eloneva, 2010; Teir et al., 2009).

According to an assessment of IEA GHG (2000), the calculated cost of HCl extraction route was €179 per tCO_2 avoided. According to Katsuyama *et al.* (2005), CO_2 sequestration by IAC of cement wastes would cost \$136–323 per $tCaCO_3$, depending on process conditions (30 bar and 50°C) and product grade. However, the costs of geological storage could lead to higher expenses (up to \$235 per tCO_2) in case of complex industrial plants (refiners, etc.) which could require non-standard solutions (IMC, 2008). Also, in case of MC, the potential leakages and prolonged monitoring costs can be avoided. In order to be feasible, MC must provide additional economic benefits in hazardous waste management (mine tailings, RM, MSWI and power plant ashes) or in production of commercial products (building materials, etc.).

So far, only a few projects have progressed to the small-scale demonstration let alone commercial phase. A 2,120 MW coal-fired power plant in Point of Rocks, USA has been implemented with a pilot scale MC unit that uses coal FA to reduce the emissions of CO_2, SO_2 and Hg (Reddy *et al.*, 2011, 2013). Examples for MC commercialisation also include APC residues-based aggregate production in UK (Gunning *et al.*, 2013) and a RM carbonation plant at Kwinana in Australia (GCCSI, 2011; IAI, 2013).

The production of artificial aggregates from CO_2 carbonation has been demonstrated by Carbon8, a spin-out company of the University of Greenwich, UK, which uses MC treatment to produce artificial aggregates from APC residues for further utilisation in the manufacturing of carbon neutral building blocks (CO_2, sand, cement) by Lignacite, UK. The global demand for the raw commodities, including cement additives, fillers and iron ore feedstock from rock and/or industrial waste mineralisation, is about 27.5 Gt and can be easily fulfilled assuming 10% abatement of the global CO_2 emissions by MC (Sanna *et al.*, 2012b). Still, the commercial viability of MC-based aggregate production has to be demonstrated on a large scale (CSLF, 2012).

Alcoa has been running RM carbonation plant since 2007 binding annually ~70 $ktCO_2$ from the neighbouring ammonia plant (GCCSI, 2011; IAI, 2013). However, the CO_2 sequestration capacity of RM is relatively low, $\sim$30 tonnes RM has to be utilised to capture $1\,tCO_2$. In order to be applicable, a concentrated ($\sim$85%) high pressure source of CO_2 source has to be located reasonably close to the alumina refinery (Jones et al., 2006; Jones and Haynes, 2011; Cooling, 2007).

For conclusion, CO_2 mineralisation by wastes could be used to improve the environmental and mechanical parameters of several wastes with CO_2 sequestration as an additional benefit. The valorisation choices closest to commercial realisation include production of shaped construction materials and precipitated $CaCO_3$. However, the CO_2 sequestration potential of wastes remains marginal in global scale of CO_2 emissions and cannot compete with underground storage. The main challenges to overcome include high processing costs, especially for process intensification (additives, ultrasound, high temperature and pressure) and low reward for the capture of CO_2 as well as waste treatment and valorisation.

References

Arickx, S., V. De Borger, T. Van Gerven and C. Vandecasteele (2010). Effect of carbonation on the leaching of organic carbon and of copper from MSWI bottom ash. *Waste Management* 30: 1296–1302.

Arickx, S., T. Van Gerven and C. Vandecasteele (2006). Accelerated carbonation for treatment of MSWI bottom ash. *Journal of Hazardous Materials* B137: 235–243.

Assima, G. P., Larachi, J. Molson and G. Beaudoin (2014a). Comparative study of five Québec ultramafic mining residues for use in direct ambient carbon dioxide mineral sequestration. *Chemical Engineering Journal* 245: 56–64.

Assima, G. P., Larachi, J. Molson and G. Beaudoin (2014b). Emulation of ambient carbon dioxide diffusion and carbonation within nickel mining residues. *Minerals Engineering* 59: 39–44.

Assima, G. P., Larachi, J. Molson and G. Beaudoin (2014c). Impact of temperature and oxygen availability on the dynamics of ambient CO_2 mineral sequestration by nickel mining residues. *Chemical Engineering Journal* 240: 394–403.

Assima, G. P., Larachi, J. Molson and G. Beaudoin (2014d). New tools for stimulating dissolution and carbonation of ultramafic mining residues. *The Canadian Journal of Chemical Engineering* 92: 2029–2038.

Baciocchi, R., G. Costa, E. D. Bartolomeo, A. Polettini and R. Pomi (2009a). The effects of accelerated carbonation on CO_2 uptake and metal release from incineration APC residues. *Waste Management* 29: 2994–3003.

Baciocchi, R., G. Costa, M. Di Gianfilippo, A. Polettini, R. Pomi and A. Stramazzo (2015). Thin-film versus slurry-phase carbonation of steel slag: CO_2 uptake and effects on mineralogy. *Journal of Hazardous Materials* 283: 302–313.

Baciocchi, R., G. Costa, E. Lategano, C. M. Polettini, R. Pomi, P. Postorino and S. Rocca (2010). Accelerated carbonation of different size fractions of bottom ash from RDF incineration. *Waste Management* 30: 1310–1317.

Baciocchi, R., G. Costa, A. Polettini, R. Pomi and V. Prigiobbe (2009b). Comparison of different reaction routes for carbonation of APC residues. *Energy Procedia* 1: 4851–4858.

Baciocchi, R., A. Polettini, R. Pomi, V. Prigiobbe, V. Zedtwitz-Nikulshyna and A. Steinfeld (2006a). Performance and kinetics of CO_2 sequestration by direct gas–solid carbonation of APC residues. In *8th International Conference of Greenhouse Gas Control Technologies*, 5 p. on CD-ROM. Trondheim, Norway: Elsevier ltd.

Baciocchi, R., A. Polettini, R. Pomi, V. Prigiobbe, V. Von Zedwitz and A. Steinfeld (2006b). CO_2 Sequestration by Direct Gas–Solid Carbonation of Air Pollution Control (APC) Residues. *Energy and Fuels* 20: 1933–1940.

Back, M., M. Kuehn, H. Stanjek and S. Peiffer (2008). Reactivity of Alkaline Lignite Fly Ashes Towards CO_2 in Water. *Environmental Science & Technology* 42: 4520–4526.

Bauer, M., N. Gassen, H. Stanjek and S. Peiffer (2011). Carbonation of lignite fly ash at ambient T and P in a semi-dry reaction system for CO_2 sequestration. *Applied Geochemistry* 26: 1502–1512.

Bertos, M. F., S. J. R. Simons, C. D. Hills and P. J. Carey (2004a). A review of accelerated carbonation technology in the treatment of cement-based materials and sequestration of CO_2. *Journal of Hazardous Materials* B112: 193–205.

Bertos, M. F., X. Li, S. J. R. Simons, C. D. Hills and P. J. Carey (2004b). Investigation of accelerated carbonation for the stabilisation of MSW incinerator ashes and the sequestration of CO_2. *Green Chemistry* 6: 428–436.

Bobicki, E. R., Q. Liu, Z. Xu and H. Zeng (2012). Carbon capture and storage using alkaline industrial wastes. *Progress in Energy and Combustion Science* 38: 302–320.

Bodor, M., R. Santos, T. Van Gerven and M. Vlad (2013). Recent developments and perspectives on the treatment of industrial wastes by mineral carbonation — a review. *Central European Journal of Engineering* 3: 566–584.

Bonenfant, D., L. Kharoune, S. Sauve, R. Hausler, P. Niquette, M. Mimeault and M. Kharoune (2008a). CO_2 Sequestration Potential of Steel Slags at Ambient Pressure and Temperature. *Industrial & Engineering Chemistry Research* 47: 7610–7616.

Bonenfant, D., L. Kharoune, S. b. Sauve, R. Hausler, P. Niquette, M. Mimeault and M. Kharoune (2008b). CO_2 Sequestration by Aqueous Red Mud Carbonation at Ambient Pressure and Temperature. *Industrial & Engineering Chemistry Research* 47: 7617–7662.

Chang, E. E., C.-H. Chen, Y.-H. Chen, S.-Y. Pan and P.-C. Chiang (2011). Performance evaluation for carbonation of steel-making slags in a slurry reactor. *Journal of Hazardous Materials* 186: 558–564.

Chang, E. E., T.-L. Chen, S.-Y. Pan, Y.-H. Chen and P.-C. Chiang (2013a). Kinetic modeling on CO_2 capture using basic oxygen furnace slag coupled with cold-rolling wastewater in a rotating packed bed. *Journal of Hazardous Materials* 260: 937–946.

Chang, E. E., A.-C. Chiu, S.-Y. Pan, Y.-H. Chen, C.-S. Tan and P.-C. Chiang (2013b). Carbonation of basic oxygen furnace slag with metalworking wastewater in a slurry reactor. *International Journal of Greenhouse Gas Control* 12: 382–389.

Chang, E. E., S.-Y. Pan, Y.-H. Chen, C.-S. Tan and P.-C. Chiang (2012). Accelerated carbonation of steelmaking slags in a high-gravity rotating packed bed. *Journal of Hazardous Materials* 227–228: 97–106.

Chiang, Y. W., R. M. Santos, J. Elsen, B. Meesschaert, J. A. Martens and T. Van Gerven (2014). Towards zero-waste mineral carbon sequestration via two-way valorization of ironmaking slag. *Chemical Engineering Journal* 249: 260–269.

Cooling, D. J. (2007). Improving the sustainability of residue management practices — Alcoa World Alumina Australia. In: *Paste 2007 — Proceedings of the 10th International Seminar on Paste and Thickened Tailings* [Fourie, A. B. and R. J. Jewell (eds.)]. Australian center for Geomechanics. Perth, Australia, 3–15.

Cornelis, G., T. Van Gerven and C. Vandecasteele (2006). Antimony leaching from uncarbonated and carbonated MSWI bottom ash. *Journal of Hazardous Materials* A137: 1284–1292.

Costa, G., R. Baciocchi, A. Polettini, R. Pomi, C. D. Hills and P. J. Carey (2007). Current status and perspectives of accelerated carbonation processes on municipal waste combustion residues. *Environmental Monitoring and Assessment* 135: 55–75.

CSLF (2012). Annual Meeting Documents Book, CSLF-T-2012-10. Perth, Australia.

Dilmore, R., P. Lu, D. Allen, Y. Soong, S. Hedges, J. K. Fu, C. L. Dobbs, A. Degalbo and C. Zhu (2007). Sequestration of CO_2 in Mixtures of Bauxite Residue and Saline Wastewater[†]. *Energy & Fuels* 22: 343–353.

Doucet, F. J. (2009). Effective CO_2-specific sequestration capacity of steel slags and variability in their leaching behaviour in view of industrial mineral carbonation. *Minerals Engineering* 23: 262–269.

Dri, M., A. Sanna and M. Maroto-Valer (2013). Dissolution of steel slag and recycled concrete aggregate in ammonium bisulphate for CO_2 mineral carbonation. *Fuel Processing Technology* 113: 114–122.

Dri, M., A. Sanna and M. Maroto-Valer (2014). Mineral carbonation from metal wastes: Effect of solid to liquid ratio on the efficiency and characterization of carbonated products. *Applied Energy* 113: 515–523.

Eloneva, S. (2010). Reduction of CO_2 emissions by mineral carbonation: steelmaking slags and raw material with a pure calcium carbonate end product. In Department of Energy Technology, 99. Helsinki, Aalto University School of Science and Technology.

Eloneva, S., A. Said, C.-J. Fogelholm and R. Zevenhoven (2012). Preliminary assessment of a method utilizing carbon dioxide and steelmaking slags to produce precipitated calcium carbonate. *Applied Energy* 90: 329–334.

Eloneva, S., S. Teir, J. Salminen, C.-J. Fogelholm and R. Zevenhoven (2008a). Fixation of CO_2 by carbonating calcium derived from blast furnace slag. *Energy* 33: 1461–1467.

Eloneva, S., S. Teir, J. Salminen, C.-J. Fogelholm and R. Zevenhoven (2008b). Steel Converter Slag as a Raw Material for Precipitation of Pure Calcium Carbonate. *Industrial & Engineering Chemistry Research* 47: 7104–7111.

Gahan, C. S., M. L. Cunha and Å. Sandström (2009). Comparative study on different steel slags as neutralising agent in bioleaching. *Hydrometallurgy* 95: 190–197.

GCCSI (2009). Strategic Analysis of the Global Status of Carbon Capture and Storage Report 2: Economic Assessment of Carbon Capture and Storage Technologies. Canberra ACT 2601 Australia: Global CCS Institute, GPO Box 828.

GCCSI (2011). Accelerating the Uptake of CCS: Industrial Use of Captured Carbon Dioxide, March 2011, Parsons Brinckerhoff and Global CCS Institute. Retrieved 20/11/2015. Available from: http://www.globalccsinstitute. com / publications / accelerating - uptake - ccs - industrial-use-captured-carbon-dioxide.

Gunning, P. J., C. D. Hills and P. J. Carey (2009). Production of lightweight aggregate from industrial waste and carbon dioxide. *Waste Management* 29: 2722–2729.

Gunning, P. J., C. D. Hills and P. J. Carey (2010). Accelerated carbonation treatment of industrial wastes. *Waste Management* 30: 1081–1090.

Gunning, P. J., C. D. Hills and P. J. Carey (2013). Commercial application of accelerated carbonation: looking back at the first year. In: *Proceedings of the fourth International Conference on Accelerated Carbonation and Materials Engineering*. Leuven, Belgium, pp. 185–192.

Han, S.-J., H. J. Im and J.-H. Wee (2015). Leaching and indirect mineral carbonation performance of coal fly ash-water solution system. *Applied Energy* 142: 274–282.

Hitch, M., S. M. Ballantyne and S. R. Hindle (2009). Revaluing mine waste rock for carbon capture and storage. *International Journal of Mining, Reclamation and Environment* 24: 64–79.

Huijgen, W. J. J. and R. N. J. Comans (2003). Carbon dioxide sequestration by mineral carbonation: Literature review. ECN-C-03016. Energy Research Centre of the Netherlands.

Huijgen, W. J. J. and R. N. J. Comans (2006). Carbonation of steel slag for Co_2 sequestration: Leaching of products and reaction mechanisms. *Environmental Science and Technology* 40: 2790–2796.

Huijgen, W. J. J., R. N. J. Comans and G.-J. Witkamp (2007). Cost evaluation of CO_2 sequestration by aqueous mineral carbonation. *Energy Conversion and Management* 48: 1923–1935.

126   *CO₂ Sequestration by Ex-Situ Mineral Carbonation*

Huijgen, W. J. J., G.-J. Witkamp and R. N. J. Comans (2005). Mineral CO_2 sequestration by steel slag carbonation. *Environmental Science and Technology* 39: 9676–9682.

Huntzinger, D. N. and T. D. Eatmon (2009). A life-cycle assessment of Portland cement manufacturing: Comparing the traditional process with alternative technologies. *Journal of Cleaner Production* 17: 668–675.

Huntzinger, D. N., J. S. Gierke, K. Kawatra, T. C. Eisele and L. L. Sutter (2009a). Carbon dioxide sequestration in cement kiln dust through mineral carbonation. *Environmental Science & Technology* 43: 1986–1992.

Huntzinger, D. N., J. S. Gierke, L. L. Sutter, S. K. Kawatrad and T. C. Eisele (2009b). Mineral carbonation for carbon sequestration in cement kiln dust from waste piles. *Journal of Hazardous Materials* 168: 31–37.

IAI (2013). Bauxite residue management: Best practice, in, International Aluminium Institute. Retrieved 11/1/2016, Available from: http://european-aluminium.eu/wp-content/uploads/2011/08/Bauxite-Residue-Management-Best-Practice_May-2013.pdf.

IEA (2015). CO_2 Emissions from Fuel Combustion. International Energy Agency. Retrieved 11/1/2016, Available from: http://www.oecd-ilibrary.org/docserver/download/6115291e.pdf?expires=1452514393&id=id&accname=oid013565&checksum=E8AF3027C0DAC327BD60783AEDE185AC.

Iizuka, A., M. Fujii, A. Yamasaki and Y. Yanagisawa (2004). Development of a New CO_2 Sequestration Process Utilizing the Carbonation of Waste Cement. *Industrial & Engineering Chemistry Research* 43: 7880–7887.

IMC (2008). Alberta CO_2 Capture Cost Survey and Supply Curve, prepared for the Alberta CCS Development Council. Ian Murray and Co. Ltd. Retrieved 20/11/2015, Available from: http://www.canadiancleanpowercoalition.com/pdf/GS25%20-%20IMC%20Report%20-%20CO₂%20Capture%20in%20Alberta.pdf.

Johnston, M., M. W. Clark, P. McMahon and N. Ward (2010). Alkalinity conversion of bauxite refinery residues by neutralization. *Journal of Hazardous Materials* 182: 710–715.

Jones, B. E. H. and R. J. Haynes (2011). Bauxite processing residue: A critical review of its formation, properties, storage, and revegetation. *Critical Reviews in Environmental Science and Technology* 41: 271–315.

Jones, G., G. Joshi, M. Clark and D. McConchie (2006). Carbon capture and the aluminium industry: Preliminary studies. *Environmental Chemistry* 3: 297–303.

Kashef-Haghighi, S. and S. Ghoshal (2009). CO_2 Sequestration in concrete through accelerated carbonation curing in a flow-through reactor. *Industrial & Engineering Chemistry Research* 49: 1143–1149.

Kashef-Haghighi, S. and S. Ghoshal (2013). Physico–chemical processes limiting Co_2 uptake in concrete during accelerated Carbonation curing. *Industrial & Engineering Chemistry Research* 52: 5529–5537.

Kashef-Haghighi, S., Y. Shao and S. Ghoshal (2015). Mathematical modeling of CO_2 uptake by concrete during accelerated carbonation curing. *Cement and Concrete Research* 67: 1–10.

Katsuyama, Y., A. Yamasaki, A. Iizuka, M. Fujii, K. Kumagai and Y. Yanagisawa (2005). Development of a process for producing high-purity calcium carbonate (CaCO3) from waste cement using pressurized CO_2. *Environmental Progress* 24: 162–170.

Ki Lee, D. (2004). An apparent kinetic model for the carbonation of calcium oxide by carbon dioxide. *Chemical Engineering Journal* 100: 71–77.

Kodama, S., T. Nishimoto, N. Yamamoto, K. Yogo and K. Yamada (2008). Development of a new pH-swing CO_2 mineralization process with a recyclable reaction solution. *Energy* 33: 776–784.

Larachi, F., I. Daldoul and G. Beaudoin (2010). Fixation of CO_2 by chrysotile in low-pressure dry and moist carbonation: *Ex-situ* and *in-situ* characterizations. *Geochimica et Cosmochimica Acta* 74: 3051–3075.

Larachi, F., J. P. Gravel, B. P. A. Grandjean and G. Beaudoin (2012). Role of steam, hydrogen and pretreatment in chrysotile gas–solid carbonation: Opportunities for pre-combustion CO_2 capture. *International Journal of Greenhouse Gas Control* 6: 69–76.

Lekakh, S. N., C. H. Rawlins, D. G. C. Robertson, V. L. Richards and K. D. Peaslee (2008). Kinetics of Aqueous Leaching and Carbonization of Steelmaking Slag. *Metallurgical and Materials Transactions B* 39: 125–134.

Li, X., M. F. Bertos, C. D. Hills, P. J. Carey and S. Simon (2007). Accelerated carbonation of municipal solid waste incineration fly ashes. *Waste Management* 27: 1200–1206.

Mattila, H.-P., H. Hudd and R. Zevenhoven (2014). Cradle-to-gate life cycle assessment of precipitated calcium carbonate production from steel converter slag. *Journal of Cleaner Production* 84: 611–618.

Mattila, H.-P. and R. Zevenhoven (2014). Design of a Continuous Process Setup for Precipitated Calcium Carbonate Production from Steel Converter Slag. *ChemSusChem* 7: 903–9013.

Mayoral, M. C., J. M. Andrés and M. P. Gimeno (2013). Optimization of mineral carbonation process for CO_2 sequestration by lime-rich coal ashes. *Fuel* 106: 448–454.

Meima, J. A., R. D. van der Weijden, T. T. Eighmy and R. N. J. Comans (2002). Carbonation processes in municipal solid waste incinerator bottom ash and their effect on the leaching of copper and molybdenum. *Applied Geochemistry* 17: 1503–1513.

Monkman, S. and Y. Shao (2006). Assessing the carbonation behavior of cementitious materials. *Journal of Materials in Civil Engineering* 18: 768–776.

Montes-Hernandez, G., R. Perez-Lopez, F. Renard, J. M. Nieto and L. Charlet (2009). Mineral sequestration of CO_2 by aqueous carbonation of coal combustion fly-ash. *Journal of Hazardous Materials* 161: 1347–1354.

Muriithi, G. N., L. F. Petrik, O. Fatoba, W. M. Gitari, F. J. Doucet, J. Nel, S. M. Nyale and P. E. Chuks (2013). Comparison of CO_2 capture by *ex-situ* accelerated carbonation and in *in-situ* naturally weathered coal fly ash. *Journal of Environmental Management* 127: 212–220.

Nyambura, M. G., G. W. Mugera, P. L. Felicia and N. P. Gathura (2011). Carbonation of brine impacted fractionated coal fly ash: Implications for CO_2 sequestration. *Journal of Environmental Management* 92: 655–644.

Olajire, A. A. (2010). CO_2 capture and separation technologies for end-of-pipe applications e A review. *Energy* 35: 2610–2628.

Olajire, A. A. (2013). A review of mineral carbonation technology in sequestration of CO_2. *Journal of Petroleum Science and Engineering* 109: 364–392.

Oskierski, H. C., B. Z. Dlugogorski and G. Jacobsen (2013). Sequestration of atmospheric CO_2 in chrysotile mine tailings of the Woodsreef Asbestos Mine, Australia: Quantitative mineralogy, isotopic fingerprinting and carbonation rates. *Chemical Geology* 358: 156–169.

Pan, S.-Y., E. E. Chang and P.-C. Chiang (2012). CO_2 capture by accelerated carbonation of alkaline wastes: A review on its principles and applications. *Aerosol and Air Quality Research* 12: 770–791.

Pan, S.-Y., P.-C. Chiang, Y.-H. Chen, C.-S. Tan and E. E. Chang (2013). *Ex Situ* CO_2 capture by carbonation of steelmaking slag coupled with metalworking wastewater in a rotating packed bed. *Environmental Science & Technology* 47: 3308–3315.

Perez-Lopez, R., G. Montes-Hernandez, J. M. Nieto, F. Renard and L. Charlet (2008). Carbonation of alkaline paper mill waste to reduce CO_2 greenhouse gas emissions into atmosphere. *Applied Geochemistry* 23: 2292–2300.

Polettini, A. and R. Pomi (2004). The leaching behavior of incinerator bottom ash as affected by accelerated ageing. *Journal of Hazardous Materials* B113: 209–215.

Reddy, K. J., S. John, H. Weber, M. D. Argyle, P. Bhattacharyya, D. T. Taylor, M. Christensen, T. Foulke and P. Fahlsing (2011). Simultaneous capture and mineralization of coal combustion flue gas carbon dioxide (CO_2). *Energy Procedia* 4: 1574–1583.

Reddy, K. J., B. Reynolds and M. D. Argyle (2013). Simultaneous capture and mineralization of coal combustion flue gas CO_2, SO_2, and Hg with fly ash particles. In: *Proceedings of the 4th International Conference on Accelerated Carbonation for Environmental and Materials Engineering*. Leuven, Belgium, pp. 175–183.

Reddy, K. J., H. Weber, P. Bhattacharyya, A. Morris, D. Taylor, M. Christensen, T. Foulke and P. Fahlsing (2010). Instantaneous Capture and Mineralization of Flue Gas Carbon Dioxide: Pilot Scale Study. Available from *Nature Proceedings*: http://dx.doi.org/10.1038/npre.2010.5404.1. Access 20 June 2016.

Rendek, E., G. Ducom and P. Germain (2006). Carbon dioxide sequestration in municipal solid waste incinerator (MSWI) bottom ash. *Journal of Hazardous Materials* B128: 73–79.

Reynolds, B., K. Reddy and M. Argyle (2014). Field Application of Accelerated Mineral Carbonation. *Minerals* 4: 191.

Ryu, K. W., S. C. Chae and Y. N. Jang (2011a). Carbonation of Chrysotile under Subcritical Conditions. *Materials Transactions* 52: 1983–1988.

Ryu, K. W., M. G. Lee and Y. N. Jang (2011b). Mechanism of tremolite carbonation. *Applied Geochemistry* 26: 1215–1221.

Sahu, R. C., R. K. Patel and B. C. Ray (2010). Neutralization of red mud using CO_2 sequestration cycle. *Journal of Hazardous Materials* 179: 29–34.

Said, A., H.-P. Mattila, M. Järvinen and R. Zevenhoven (2013). Production of precipitated calcium carbonate (PCC) from steelmaking slag for fixation of CO_2. *Applied Energy* 112: 765–771.

Said, A., O. Mattila, S. Eloneva and M. Järvinen (2015). Enhancement of calcium dissolution from steel slag by ultrasound. *Chemical Engineering and Processing: Process Intensification* 89: 1–8.

Sanna, A., M. Dri, M. R. Hall and M. Maroto-Valer (2012a). Waste materials for carbon capture and storage by mineralisation (CCSM) — A UK perspective. *Applied Energy* 99: 545–554.

Sanna, A., M. R. Hall and M. Maroto-Valer (2012b). Post-processing pathways in carbon capture and storage by mineral carbonation (CCSM) towards the introduction of carbon neutral materials. *Energy & Environmental Science* 5: 7781–7796.

Sanna, A., M. Uibu, G. Caramanna, R. Kuusik and M. Maroto-Valer (2014). A review of mineral carbonation technologies to sequester CO_2. *Chemical Society Reviews* 43: 8049–8080.

Santos, R. M., M. Bodor, P. N. Dragomir, A. G. Vraciu, M. Vlad and T. Van Gerven (2014). Magnesium chloride as a leaching and aragonite-promoting self-regenerative additive for the mineral carbonation of calcium-rich materials. *Minerals Engineering* 59: 71–81.

Santos, R. M., D. François, G. Mertens, J. Elsen and T. Van Gerven (2013a). Ultrasound-intensified mineral carbonation. *Applied Thermal Engineering* 57: 154–163.

Santos, R. M., D. Ling, A. Sarvaramini, M. Guo, J. Elsen, F. Larachi, G. Beaudoin, B. Blanpain and T. Van Gerven (2012). Stabilization of basic oxygen furnace slag by hot-stage carbonation treatment. *Chemical Engineering Journal* 203: 239–250.

Santos, R. M., G. Mertens, M. Salman, Ö. Cizer and T. Van Gerven (2013b). Comparative study of ageing, heat treatment and accelerated carbonation for stabilization of municipal solid waste incineration bottom ash in view of reducing regulated heavy metal/metalloid leaching. *Journal of Environmental Management* 128: 807–821.

Santos, R. M., J. Van Bouwel, E. Vandevelde, G. Mertens, J. Elsen and T. Van Gerven (2013c). Accelerated mineral carbonation of stainless steel slags for CO_2 storage and waste valorization: Effect of process parameters on geochemical properties. *International Journal of Greenhouse Gas Control* 17: 32–45.

Shih, S.-M., C. U.-S. Ho, Y.-S. Song and J.-P. Lin (1999). Kinetics of the reaction of Ca(OH)2 with CO_2 at low temperature. *Industrial & Engineering Chemistry Research* 38: 1316–1322.

Shuto, D., K. Igarashi, H. Nagasawa, A. Iizuka, M. Inoue, M. Noguchi and A. Yamasaki (2015). CO_2 fixation process with waste cement powder via regeneration of alkali and acid by electrodialysis: Effect of operation conditions. *Industrial & Engineering Chemistry Research* 54: 6569–6577.

Shuto, D., H. Nagasawa, A. Iizuka and A. Yamasaki (2014). A CO_2 fixation process with waste cement powder via regeneration of alkali and acid by electrodialysis. *RSC Advances* 4: 19778–19788.

Si, C., Y. Ma and C. Lin (2013). Red mud as a carbon sink: Variability, affecting factors and environmental significance. *Journal of Hazardous Materials* 244–245: 54–59.

Soong, Y., D. L. Fauth, B. H. Howard, B. H., R. J. Jones, D. K. Harrison, A. L. Goodman, M. L. Gray and E. A. Frommell (2005). CO_2 sequestration with brine solution and fly ashes. *Energy Conversion and Management* 47: 1676–1685.

SpecialChem (2013). Manufacturing Process e Precipitated Calcium Carbonate Center. Available from: http://polymer-additives.specialchem.com/ selection-guide/precipitated-calcium-carbonate-center/cristallinities. Access 20 November 2015.

Stolaroff, J. K., G. V. Lowry and D. W. Keith (2005). Using CaO- and MgO-rich industrial waste streams for carbon sequestration. *Energy Conversion and Management* 46: 687–699.

Sun, J., M. F. Bertos and S. J. R. Simons (2008). Kinetic study of accelerated carbonation of municipal solid waste incinerator air pollution control residues for sequestration of flue gas CO_2. *Energy and Environmental Science* 1: 370–377.

Teir, S., S. Eloneva, C.-J. Fogelholm and R. Zevenhoven (2007a). Dissolution of steelmaking slags in acetic acid for precipitated calcium carbonate production. *Energy* 32: 528–539.

Teir, S., S. Eloneva, C.-J. Fogelholm and R. Zevenhoven (2009). Fixation of carbon dioxide by producing hydromagnesite from serpentinite. *Applied Energy* 86: 214–218.

Teir, S., R. Kuusik, C.-J. Fogelholm and R. Zevenhoven (2007b). Production of magnesium carbonates from serpentinite for long-term storage of CO_2. *International Journal of Mineral Processing* 85: 1–15.

Teramura, S., N. Isu and K. Inagaki (2000). New Building Material from Waste Concrete by Carbonation. *Journal of Materials in Civil Engineering* 12: 288–293.

Uibu, M. and R. Kuusik (2009). Mineral trapping of CO_2 via oil shale ash aqueous carbonation: process rate controlling mechanism and developments of continuous mode reactor system. *Oil Shale* 26: 40–58.

Uibu, M., R. Kuusik, L. Andreas and K. Kirsimäe (2011). The CO_2 -binding by Ca-Mg-silicates in direct aqueous carbonation of oil shale ash and steel slag. *Energy Procedia* 4: 925–932.

Uibu, M., M. Uus and R. Kuusik (2009). CO_2 mineral sequestration in oil-shale wastes from Estonian power production. *Journal of Environmental Management* 90: 1253–1260.

Uibu, M., O. Velts and R. Kuusik (2010). Developments in CO_2 mineral carbonation of oil shale ash. *Journal of Hazardous Materials* 174: 209–214.

Uliasz-Bocheńczyk, A., E. Mokrzycki, Z. Piotrowski and R. Pomykał (2009). Estimation of CO_2 sequestration potential via mineral carbonation in fly ash from lignite combustion in Poland. *Energy Procedia* 1: 4873–4879.

Van Gerven, T., E. Van Keer, S. Arickx, M. Jaspers, G. Wauters and C. Vandecasteele (2005). Carbonation of MSWI-bottom ash to decrease heavy metal leaching, in view of recycling. *Waste Management* 25: 291–300.

Velts, O., M. Uibu, J. Kallas and R. Kuusik (2011). Waste oil shale ash as a novel source of calcium for precipitated calcium carbonate: Carbonation mechanism, modeling, and product characterization. *Journal of Hazardous Materials* 195: 139–146.

Velts, O., M. Kindsigo, M. Uibu, J. Kallas and R. Kuusik (2014). CO_2 mineralisation: production of CaCO3-type material in a continuous flow disintegrator-reactor. *Energy Procedia* 63: 5904–5911.

Wang, L., Y. Jin and Y. Nie (2010). Investigation of accelerated and natural carbonation of MSWI fly ash with a high content of Ca. *Journal of Hazardous Materials* 174: 334–343.

Wee, J.-H. (2013). A review on carbon dioxide capture and storage technology using coal fly ash. *Applied Energy* 106: 143–151.

Wilson, S. A., G. M. Dipple, I. M. Power, J. M. Thom, R. G. Anderson, M. Raudsepp, J. E. Gabites and G. Southam (2009). Carbon dioxide fixation within mine wastes of ultramafic-hosted ore deposits: Examples from the Clinton creek and cassiar chrysotile deposits, Canada. *Economic Geology* 104: 95–112.

Yadav, V. S., M. Prasad, J. Khan, S. S. Amritphale, M. Singh and C. B. Raju (2010). Sequestration of carbon dioxide (CO_2) using red mud. *Journal of Hazardous Materials* 176: 1044–1050.

Zingaretti, D., G. Costa and R. Baciocchi (2013). Assessment of the Energy Requirements for CO_2 Storage by Carbonation of Industrial Residues. Part 1: Definition of the Process Layout. *Energy Procedia* 37: 5850–5857.

Chapter 5

MC Process Scale and Product Applications

Tze Yuen Yeo and Jie Bu
*Institute of Chemical and Engineering Sciences (ICES),
A*STAR, Singapore*

5.1 Introduction

Climate scientists generally agree that unabated business-as-usual emissions of CO_2 will lead to "further warming and long-lasting changes in all components of the climate system, increasing the likelihood of severe, pervasive and irreversible impacts for people and ecosystems" (Pachauri *et al.*, 2014). To prevent this, global warming should be limited to 2°C from pre-industrial atmospheric temperatures (Anderson and Bows, 2011).

In order to meet this requirement, the International Energy Agency (IEA) estimates that 120 Gt of CO_2 will have to be captured and sequestered between 2015 and 2050 (Levina *et al.*, 2013). This means that on average, 3.5 BT of CO_2 must be prevented from being emitted into the atmosphere every year. With the various tools that we currently have, this represents a daunting, but nevertheless doable task. Achieving this would require a combined effort that includes improvements in energy efficiency, greener manufacturing processes and carbon capture and storage/utilisation (CCSU) technologies across all sectors of industry.

Mineral carbonation (MC) is one of the several technological options for CCSU. Compared with the other alternatives, MC may prove to be a very effective CCSU technology simply because it is perfectly suited to handle the large amounts of CO_2 that has to be sequestered (Sanna *et al.*, 2012). However, the utilisation or disposal of the carbonate products is often a concern for policymakers when considering MC as a CCSU technology. These concerns are addressed in this chapter, and it will be shown that even with large-scale deployment of MC, the product quantities are not impossible to handle and dispose of.

For example, consider the conversion of one Gt of CO_2 into a carbonate/silica mixture using a pressure carbonation process. From 1 Gt of CO_2, approximately 5 Gt of a carbonate/silica mixture is obtained (equivalent to about 2.5 billion cubic meters of material). Nearly 72 km^2 of land can be reclaimed from marine environments, assuming the surrounding seas are 35 m deep. This is equivalent to merely one-tenth of the total land area of Singapore, a small city state where land is scarce. Further in the chapter, we discuss various other applications and utilisation pathways of the products of MC, and it will be seen that there are a variety of downstream applications for the products. In other words, utilisation and disposal of the MC products are not a problem in many cases.

Like all industrial processes, if MC process technologies are to be widely implemented on a commercial scale, they have to satisfy several criteria. The practicality, economics and energy consumption of the process, as well as quality, quantity and marketability of the products have to be taken into consideration.

Of these considerations, process economics is often the main determining factor whether a process is developed beyond pilot-scale operations. In other words, a high profit margin often serves as a good motivating force for the development and deployment of process technologies. For example, it is not a coincidence that enhanced oil recovery (EOR) currently appears to be the largest and most widely employed carbon sequestration method in the world. Apart from storing captured CO_2, it also yields a marginal benefit from increased oil production. This revenue-generating aspect of EOR makes it easier for policymakers to adopt and encourage.

MC as a whole can therefore learn from the relative success of EOR. In order to succeed, it has to prove itself as a profitable (or at least almost cost-neutral) technology.

MC also stands out compared to other CCSU options, due to its inherent strengths in the following areas:

(1) **Permanence:** Carbon dioxide sequestered as mineral carbonates do not require constant monitoring to prevent re-emission into the atmosphere.

(2) **Scalability:** MC is likely the only sequestration method that can match and accommodate the vast amounts of carbon dioxide that needs to be removed.

(3) **Versatility:** The conversion of carbon dioxide into mineral carbonates can be achieved through many ways, and each MC process can be tailored to adapt to individual needs and situations.

MC process technologies will have to take advantage of these strengths, and translate them into economic benefits in order to thrive and succeed as industrially relevant processes.

This chapter is focused on discussing the above three aspects of MC, and how these aspects can be expanded upon to strengthen the case for the development and widespread implementation of MC. More specifically, the discussions that follow will first outline the alternatives to MC, and show how they are inadequate to permanently contribute to meaningful CO_2 reductions. Then, the scale of MC and its relation to the downstream markets for its products is described in detail. These descriptions will attempt to familiarise the reader with the potential applications and approximate demand for the products, and in the process attempt to identify the existing industries that MC can co-exist and form a synergistic relationship with. Finally, a hypothetical hybrid process comprising a pressure carbonation segment and a pH-swing segment is discussed, showing how the two segments can complement each other by bringing mutual benefits to the overall integrated process.

5.2 Current CCSU Alternatives to MC

In this section, we begin the discussion with brief descriptions of the CCSU alternatives to MC. These alternative technologies are then used as a basis for comparison in the discussion of MC in general.

Figure 5.1 shows the capacities of various CO_2 storage and utilisation methods. The chart is split into two sections, differentiating between methods that involve transforming CO_2 into a different chemical species before sequestration (left half) and those that store or use CO_2 without chemical transformation (right half). It can be seen that of all the currently proposed methods for CCSU, only MC can permanently and safely sequester the large quantities of CO_2 that are required to prevent the earth from warming beyond 2°C.

5.2.1 *Storage/Utilisation Methods without Chemical Transformation of CO_2*

Methods that do not require chemical transformation of CO_2 typically include the *direct utilisation* of the gas in various industrial

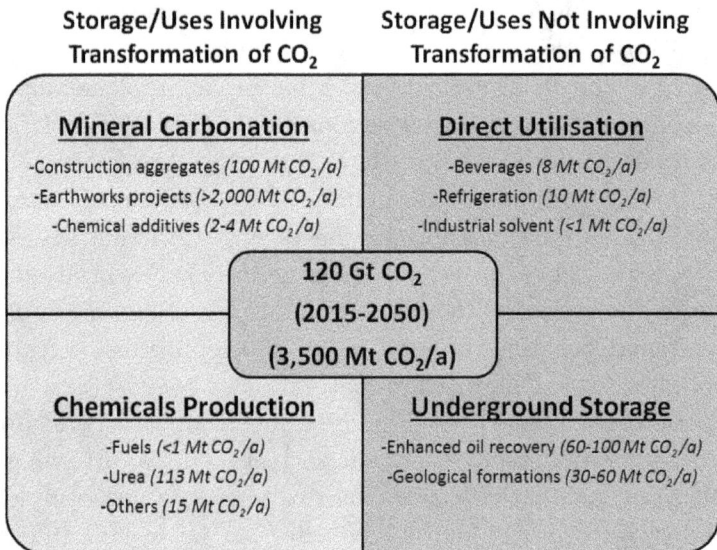

Storage/Uses Involving Transformation of CO_2	Storage/Uses Not Involving Transformation of CO_2
Mineral Carbonation	**Direct Utilisation**
-Construction aggregates *(100 Mt CO_2/a)*	-Beverages *(8 Mt CO_2/a)*
-Earthworks projects *(>2,000 Mt CO_2/a)*	-Refrigeration *(10 Mt CO_2/a)*
-Chemical additives *(2-4 Mt CO_2/a)*	-Industrial solvent *(<1 Mt CO_2/a)*
120 Gt CO_2 (2015-2050) (3,500 Mt CO_2/a)	
Chemicals Production	**Underground Storage**
-Fuels *(<1 Mt CO_2/a)*	-Enhanced oil recovery *(60-100 Mt CO_2/a)*
-Urea *(113 Mt CO_2/a)*	-Geological formations *(30-60 Mt CO_2/a)*
-Others *(15 Mt CO_2/a)*	

Figure 5.1 Scale and capacities of various storage/utilisation methods for CO_2 sequestration.

processes, such as in the manufacture of carbonated beverages or fire extinguishers. CO_2 can also be used as a refrigerant (dry ice) or solvent (supercritical CO_2 for decaffeination of coffee beans, for example).

However, currently the largest direct use of CO_2 involves compressing it into the supercritical state and used for *EOR*. Approximately 60–100 Mt of CO_2 per year (Tao and Clarens, 2013) are injected into oil wells to enhance the recovery of crude oil, as well as storing the CO_2 in the vacated pores left behind after oil extraction (Jaramillo *et al.*, 2009). Supercritical CO_2 can also be stored directly in *underground geological formations*, especially in between layers of non-permeable rock where it can effectively be sealed and prevented from escaping (Hendriks and Blok, 1993).

It is obvious that the sequestration of CO_2 must involve a certain degree of permanence in its storage; otherwise it will then become an exercise in futility. Most of the direct utilisation methods will eventually re-release the CO_2 into the atmosphere, making them temporary carbon sinks at best. EOR and underground storage offer much larger capacities as well as longer sequestration lifetimes, but these remain susceptible to re-emissions of the stored CO_2 and require constant monitoring to ensure it stays underground (Streit and Hillis, 2004).

5.2.2 *Storage/Utilisation Methods Involving Chemical Transformation of* CO_2

Captured CO_2 can also be converted into other chemicals. A lot of research focuses on converting CO_2 into *carbon-based fuel*, thus in theory essentially closing the carbon cycle (Liu *et al.*, 2003). However, the energy content of the fuel produced from CO_2 will always be less than the energy input of the conversion process (Taheri Najafabadi, 2013). Thus, a caveat exists that the energy input for producing the fuel must be obtained from non-fossil sources. Despite the constant development of more efficient catalysts for CO_2-to-fuel conversion, these processes remain confined to lab scale or demonstration scale.

Apart from fuel, CO_2 can also be used as a basic building block or raw material for production of other chemicals. For example, the production of *urea* consumes large quantities of CO_2 every year (Meessen, 2000):

$$2NH_3 + CO_2 \rightarrow (NH_2)_2CO + H_2O \qquad (5.1)$$

Though mainly used as a fertiliser, urea is a versatile chemical that finds a variety of uses across many industries (Mavrovic *et al.*, 2000).

Recent developments in polymer synthesis have also led to increasing use of CO_2 as a replacement for toxic phosgene gas in the manufacture of *polycarbonate materials*. Polycarbonates are copolymers comprising alternating monomers of bisphenols and CO_2, as shown in Figure 5.2.

It can be seen that CO_2 takes up only a small weight percentage of the polycarbonate material (around 17% in the case of bisphenol A co-monomer). In other words, for every tonne of polycarbonate produced, only 0.17 tonne of CO_2 are used in the manufacturing process. For comparison, the demand for polycarbonates in 2013 was approximately 3.6 Mt, which translates to about 612 kt of CO_2 required as raw materials (Abts *et al.*, 2000).

While the utilisation of CO_2 via transformation into other chemicals might seem promising, we must consider the life cycle of these products that utilise CO_2 in their manufacture. In the case of urea, most of the CO_2 that is consumed is obtained as a by-product of ammonia production (Wood and Cowie, 2004). Therefore, it is very unlikely that the production of urea can contribute significantly

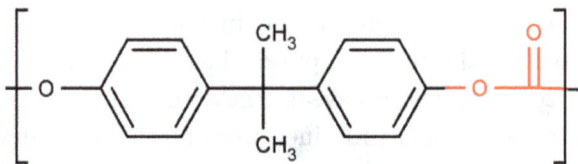

Figure 5.2 Structure of polycarbonate monomer units comprising bisphenol A (black) and CO_2 (red).

towards reducing CO_2 generated from other point sources other than ammonia production.[1]

Furthermore, plastics such as polycarbonates typically end up in landfills or incinerators at the end of their life, resulting in pollution or re-emission of the incorporated CO_2 into the atmosphere. As such, it can be argued that plastics are not a viable or large enough carbon sink for the sequestration of CO_2.

5.3 The Scale and Permanence of MC

Next, we consider the two most important criteria for CO_2 sequestration: scale and permanence. MC converts gaseous carbon dioxide into solid inorganic carbonates by binding them to alkaline-earth metal oxides. These inorganic carbonates are exceptionally stable at ambient conditions, decomposing and releasing the bound CO_2 only at high temperatures (Hemmati *et al.*, 2014). Mineral carbonates are also chemically inert, as they only susceptible to attack at acidic pH values. Therefore, in other words, once carbon dioxide is converted into an alkaline-earth carbonate, it can for all intents and purposes be considered permanently and safely sequestered. However, this also means that several traditional and large scale uses for the carbonate product have to be ruled out, since these uses often entail the re-emission of the bound CO_2. Examples of these uses include the production of refractory materials, cement and agricultural lime.

MC also requires a large and constant supply of alkalinity, mainly in the form of alkaline-earth metals such as MgO and CaO. Fortunately, these elements are in abundant supply in our

---

[1]The EPA estimates that 8 Mt of ammonia were produced in the United States in 2006. The associated CO_2 emissions were approximately 14.5 Mt, which is a carbon footprint of 1.8 tonnes of CO_2 per tonne ammonia produced. For urea production, 1 tonne of ammonia reacts with 1.3 tonnes of CO_2 to give about 1.76 tonnes of urea. This entails a net CO_2 emission of 0.5 tonne ammonia produced, even if the entire world's ammonia supply were used for urea production.

planet's crust. For example, on a molar basis, magnesium atoms are approximately 50 times more abundant than carbon atoms on earth (Railsback, 2015). Thus, in theory it should be possible to produce enough MgO to sequester all of the carbon dioxide that can be produced.

Although alkaline-earth elements might be plentiful in the earth's crust, they must be concentrated enough to be easily extractable for MC to become a feasible process technology. Recent surveys have reported large deposits of magnesium-silicate minerals in various locations around the world, including one massive ophiolite formation in Oman that is estimated to be able to sequester up to 77 trillion tonnes (77,000 Gt) of CO_2 (Kelemen and Matter, 2008). While it quickly becomes apparent that not all the minerals can and will be used for MC, formations like this nevertheless represent considerable carbon sinks. Apart from minerals, magnesium and calcium salts can also be extracted from brines and seawater and used as an alternative source of alkalinity. At concentrations of approximately 1,300 ppm, the world's oceans represent a practically inexhaustible source of magnesium ions for MC (Shand, 2006).

On the other hand, all of the known and listed reserves of fossil fuels would emit approximately 745 Gt of CO_2 if they were consumed entirely (Leaton, 2012). This amount is two to three orders of magnitude smaller compared to the estimated sequestration capacity of worldwide deposits of magnesium-silicate minerals. Thus, in both theory and practice, the sources of alkaline-earth minerals for MC should be more than able to match the scale of modern CO_2 emissions.

However, unlike other CCSU technologies, it is impossible to specify in an exact manner the mass balance for any given MC process technology due to the variation in mineral compositions across different localities. Even when a process is very well defined (such that typical process conditions and yields are known), small variations in mineral feed can affect the mass balance and overall process efficiency. Nevertheless, serpentine typically contains 25–40% MgO, with the rest made up of silica, various transition metal oxides and crystalline water (Uehara and Shirozu, 1985). Thus, depending on the quality of the serpentine mineral, it can take anywhere from 2 to 4 tonnes of serpentine mineral to sequester 1 tonne of CO_2 as mineral carbonates.

A quick, back-of-the-envelope calculation can be used to estimate the overall size of MC and contrast it with other industries with similarly large material flows. If we assume a worst-case scenario where 4 tonnes of raw mineral are required to sequester 1 tonne of captured CO_2, 14 Gt/y of serpentine would be required to convert 3.5 Gt/y of CO_2 into mineral carbonates (a ratio of 4 tonnes of mineral per tonne CO_2 captured). This results in the annual production of 17.5 Gt of carbonate and silica material, which would then have to be utilised or disposed of, mainly in landfills or earthwork projects such as land reclamation.

In terms of the scale of solid materials being handled, the closest analogues would perhaps be the mining and construction industries. In 2014, the construction industry consumed 4.2 Gt of cement to produce approximately 42 Gt of concrete (USGS, 2015a). Since typical concrete compositions consist of about 10% cement, 15% water and 75% aggregates, this would mean that the demand for aggregates for construction amounts to roughly 31.5 Gt/y (PCA, 2015). This is 13 Gt more than the carbonate and silica materials that will be produced from MC annually, and if fully sourced from the solid products of MC, translates to about 5–7 Gt CO_2 captured and sequestered per year (compare to the projected 3.5 Gt/y CCSU requirements).

In the same year, the United States produced nearly 1.3 Gt of crushed stone, with limestone constituting the majority of the stone mined (USGS, 2015b). According to USGS surveys,[2] nearly 500 Mt of the crushed stone were used as road paving and construction materials, implying a fairly large market for non-calcined uses of aggregates. Bearing in mind that these statistics are only limited to applications in the US, it is safe to assume that the worldwide

---

[2]Based on self-reported data. Only 46% of materials were declared to have specific uses. 27% of the materials did not have specified end uses, and another 27% went to non-respondents.

demand for these aggregates as road and construction material would easily be at least double or triple that amount.[3]

Compared with construction aggregate, land reclamation projects represent a much smaller, but still significant market for aggregates and sand. The solid products from MC can replace the sand demand for land reclamation uses, therefore avoiding ecological harm due to excessive dredging of beaches and other sand resources.

Although primarily consisting of alkaline-earth oxides and silica, iron oxides are also present in significant quantities in the raw minerals. Assuming a typical Fe_3O_4 content of 10% in the raw mineral, 14 Gt/y of serpentine (enough to capture 3.5 Gt CO_2/y) can potentially yield up to 1.4 Gt/y of iron ore by applying the appropriate separation techniques during the mineral pre-processing steps. For comparison, the total world production of iron ore from mines in 2013 was approximately 3.2 Gt (USGS, 2015c).

As such, it can be seen that the products of MC can be gradually eased into various industries (as substitutes for current sources of materials) without significantly affecting their market supply and demand balances.

5.4 Applications of MC Products

Magnesium or calcium carbonates, silicon dioxide and iron oxides or hydroxides are the primary products of MC. Depending on the chosen process technology, these materials can occur as mixtures or relatively pure compounds. Typically, pressure carbonation processes yield mixtures of all three materials (Sanna *et al.*, 2014). These processes also tend to be cheaper due to their simplicity, but the product is often less valuable and is limited to non-specialised, bulk applications. On the other hand, aqueous pH-swing processes

---

[3]The worldwide demand for bulk materials can be extrapolated from the consumption figures in the US and China, with the assumption that the US accounts for one quarter to a third of the total world demand, and China accounts for approximately two-fifths to one half. It must be emphasised that this is a highly imprecise estimation, but this assumption is only intended to give order-of-magnitude estimations for bulk materials for comparison with mineral carbonation as a whole.

Table 5.1   Selected fine and bulk applications of products from MC.

	Fine Applications	Bulk Applications
Mg/Ca Carbonates	Paper brightening coatings	Construction aggregate
	Filler material for polymers	
	Porous filtration coatings	Land reclamation/landfill
	Soda ash manufacture	
Silica	Glass	
	Tire additives	
	Refractory materials	
Iron Oxides/ Hydroxides	Pigments	Iron/steel manufacture
	Thermite	
	Magnetic devices	
Others	Stainless steel (Cr, Ni, etc.)	
	Pigments	

allow for separation of the alkaline-earth carbonates, silica and iron compounds (Sanna *et al.*, 2013). These processes are more complicated and thus usually more expensive, but they are capable of producing useful products that can serve as raw materials for a wider variety of industries.

Table 5.1 shows a list of potential uses for the products from MC. Fine applications of these materials often require reasonably pure products, while bulk applications are usually less stringent in terms of quality. End uses such as soil acidity regulators (agricultural lime), cement manufacture and refractory material (from carbonates) are not considered, as these involve re-releasing the bound CO_2 into the atmosphere.

The following subsections discuss the bulk and fine applications of these materials; the discussions will attempt to give order-of-magnitude estimates of the market sizes for these materials. Approximate prices of the products are also given whenever possible. These numbers on market size and pricing are intentionally imprecise to take into account the limits on information and fluctuations in short- to medium-term market conditions. Any significant change in market conditions or process technology would also affect the prices of bulk products, effectively making the prices given here outdated.

The market sizes are only given as an approximation, since the purpose of citing these sizes is not to make a detailed business case for MC, but to compare the markets to the overall scale of MC. Some discrepancy in market sizes between reports is to be expected, but these are generally irrelevant to the overall big picture. For example, a seemingly huge difference of 5–10 Mt in the reported market size of some aggregate products is probably still within the error margin for estimates, especially when compared with the gigatonne scale of MC. However, in spite of the imprecision in market size and pricing, the numbers given here should be fairly indicative of the order of magnitude of the demand for these products. In other words, the numbers given here are intended to serve as a guide to identify possible industries that can form a symbiotic relationship with MC.

5.4.1 *Bulk Applications of MC Products*

The carbonate/silica mixture from pressure carbonation can act as a substitute for conventional earth, sand and soil materials. For this to work, the mixture must be substantially free from toxic heavy metals that may leach out into the surrounding waters. Raw serpentine mineral contains significant amounts of nickel and chromium, and the flora and fauna on its soils are typically less biologically diverse due to the associated harsh conditions (Brady *et al.*, 2005). Therefore, pre-treating the carbonate mixture to remove heavy metals will be necessary to avoid contamination of the natural environment (Sprung and Kropp, 2000). This pre-treatment step is ideally conducted prior to the carbonation step, when it is much easier to do so. The same precaution also applies even if the carbonate/silica mixture is used to reclaim mines or as a construction aggregate, where the heavy metals can be leached by rainwater into the environment.

Nevertheless, land reclamation from aquatic environments is expected to become an increasingly important adaptation tool, as a result of the effects of rising sea levels and more intense storms due to climate change (Stern, 2006). Many low lying areas, such as New Orleans, Venice and Amsterdam are often threatened either by storms, tides or seawaters. These places can benefit from the construction of levees, dams and seawalls, which can be built quickly and easily simply by piling solid materials to form a low-tech but effective physical barrier.

Application of the carbonate/silica mixture from MC for these earthwork projects can help to reduce the dependence on excessive mining activities. The Mississippi river levee system is an impressive construction project that goes back for centuries, involving more than 5,600 km of levees that are several metres in height (US Army, 2015). While the amount of sand and soil used in its construction will probably never be precisely known, it can be safely assumed that massive quantities of material have gone into its construction, not to mention the reparations and extensions over its centuries of service. It is entirely conceivable that similar systems can be engineered in areas prone to flooding, such as along the Yellow and Ganges rivers (Chen *et al.*, 2012).

Apart from levees and dams, the carbonate/silica mixture can also be a useful material for reclaiming land from coastal regions. Like the Mississippi river levee system, large quantities of sand and other solid materials are required for the reclamation of land from seas. Since the land being reclaimed is typically close to shore, it often sits on the continental shelf region of the ocean where the seas usually average around 30–40 m in depth, often going to depths up to 200 m. This would imply that on average, 35 Mm^3 (or about 70 Mt)[4] of solids are used for every square kilometre of new land created.

For illustration purposes, consider the case of Singapore's land reclamation projects over the past few decades. Since the 1960s, Singapore has added nearly 140 km^2 to its land area, using imported sand to reclaim land from its coastal waters (Singstat, 2014). Since the average depth of the coastal waters in Singapore is around 35 m, this means that (to date) nearly 10 BT of sand may have been used for the reclamation projects (Tkalich *et al.*, 2013). The equivalent amount of the carbonate/silica mixture would have contained around

---

[4]This is calculated by assuming a material density of approximately 2,000 kg/m^3 for the carbonate/silica mixture, similar to marl (a mixture of calcium carbonate and clay).

2 Gt of CO_2, or nearly 60% of the annual amount of CO_2 to be captured and stored.

The economic argument in favour of using the carbonate/silica mixture for land reclamation is also very compelling. Many land reclamation projects are often undertaken in places where land is scarce (for example, Singapore, Hong Kong and Japan), waterlogged and unusable (for example, the Netherlands) or intended for high-end development (for example, Dubai, Monaco and Macau). As a result, the reclaimed land is typically highly valuable, often fetching high prices when sold for development. The revenue obtained from sales of reclaimed land can thus indirectly incentivise the implementation of MC, provided the right policies are put in place to support these activities.

On the other hand, concrete production appears to be a large market for the disposal of the carbonate/silica mixture at first glance. However, despite the apparent potential to absorb the large volumes of aggregate products produced by MC, concrete production is ill-equipped to accommodate the carbonate/silica mixture.

The properties of concrete are heavily dependent on the nature of the aggregates used, and are therefore sensitive to changes in particle size, impurities and composition of the aggregate raw materials (Kosmatka and Wilson, 2011). For the carbonate/silica mixture to be used widely as construction aggregate, the mixture has to adhere to some very stringent standards (civil engineers tend to be conservative and prefer to err on the side of caution when it comes to materials of construction). These standards are typically very clearly defined. For example, ASTM International has published several guidelines recommending the qualities of aggregates used in different types of concretes.

The carbonate/silica mixture is unsuitable for concrete production because of the alkali-carbonate reaction (ACR) (Swenson and Gillott, 1964). The ACR, as the name implies, occurs when carbonate components present in the aggregate reacts with the alkali contained in the cement binder:

$$MgCO_3 + 2\,NaOH \rightarrow Mg(OH)_2 + Na_2CO_3 \qquad (5.2)$$

$Mg(OH)_2$ is formed when ACR occurs, which then absorbs water from the environment. The concrete expands and fractures upon absorption of water, resulting in loss of strength and degradation. As a result, ASTM generally recommends avoiding the use of reactive carbonate rocks as aggregate for concrete production (ASTM, 2004). Certain processes (for example the NETL pressure carbonation process) may also use additives that contain a significant amount of chlorides, and this may also contribute to corrosion of steel rebars and further weakening the concrete over long periods of time (Glass and Buenfeld, 1997). In other words, judging from the likely chemical composition of the carbonate/silica mixture, it is exceedingly unlikely that it is going to be suitable for use as concrete aggregate.

Apart from the chemical composition, particle sizes also play a role in determining whether an aggregate is suitable for concrete production or not. ASTM standards recommend the following grading requirements for fine aggregate:

As can be seen from Table 5.2, particles smaller than 150 μm are generally undesirable, due to their deleterious effects on concrete strength after setting. In many instances of pressure carbonation processes, intensive grinding of the raw serpentinite mineral is required to increase Mg or Ca extraction to enhance the conversion of CO_2 to carbonates. The resulting carbonate/silica mixtures thus consist of very fine particles, often much smaller than 75 μm. Again, this is

Table 5.2 Grading requirements for fine aggregate.

Sieve	Percent Passing
9.5 mm (3/8 inch)	100
4.75 mm (No. 4)	95–100
2.36 mm (No. 8)	80–100
1.18 mm (No. 16)	50–85
600 μm (No. 30)	25–60
300 μm (No. 50)	5–30
150 μm (No. 100)	0–10
75 μm (No. 200)	0–3

a drawback when weighing the pros and cons of the carbonate/silica mixture as concrete aggregate.

Although the carbonate/silica mixture is unsuitable for use as concrete aggregate, it may find other large-scale uses for structural purposes. Asphalt paving is one such application. In 2007, approximately 1.6 Gt of asphalt material were produced, which were mainly used for laying roads and pavements (National Asphalt Paving Association, 2011). Although asphalt compositions may vary according to individual needs, it is typically a mix of 95% aggregate and 5% bitumen (Speight, 2000). Thus, it can be inferred that in 2007, approximately 1.5 Gt of aggregate was required for asphalt production. While much smaller than the required material tonnage for MC (about 10 times smaller than the 17.5 Gt/y required), it nevertheless represents a major disposal route for the carbonate/silica mixture from pressure carbonation.

In terms of chemical composition of the aggregates, asphalt manufacturing is less demanding than concrete. A variety of material types can be used as aggregate for asphalt, and these include ground glass, rubber, limestone ($CaCO_3$) and even recycled asphalt (Miliutenko *et al.*, 2013). Since asphalt is laid by mixing bitumen and aggregate at around 160°C and then spread out over a road surface, there is no risk of thermal decomposition of the carbonate material to release the bound CO_2 (recall that the bound CO_2 is released only at temperatures in excess of 350°C). Also, decades of experience in using limestone in asphalt mixes has demonstrated the feasibility of using carbonate materials in asphalt manufacturing.

The only limiting factor for the use of the carbonate/silica mixture in asphalt appears to be the particle size. As with concrete, there is usually an optimum range of aggregate sizes in asphalt for various applications. The aggregates used in road asphalt are typically sized at around 1 cm in diameter, with other sizes also being used depending on the specific applications. Like concrete, the aggregate size also affects the properties of asphalt (Colbert and You, 2012). While particles less than 75 μm in diameter can be used as aggregate in asphalt, these are often used as space filler between coarser materials.

Assuming that the coarse aggregates in asphalt observe a random close packing behaviour, around 25–30% of the remaining volume in asphalt would be occupied by fine aggregates.[5] This would correspond to a worldwide demand of around 350–450 Mt per year of fine aggregate, which is considerably smaller than the total MC scale. Perhaps improvements in process conditions to allow larger particle sizes in reaction would improve the prospects for using the carbonate/silica mixture in asphalt production.

Serpentine also typically contains about 5–15 wt.% of iron oxides, which can be extracted and sold separately as an additional source of income for MC operations. Depending on the age and weathering conditions of the mineral ore, the type of iron oxide found in serpentine can vary in oxidation state and morphology. In younger ores, the iron oxides usually exist as magnetite (Fe_3O_4), a mixture of Fe^{2+} and Fe^{3+} species (Caillaud *et al.*, 2006). Magnetite can easily be separated from serpentine via physical extraction methods such as grinding followed by magnetic separation or flotation. Older minerals will contain a larger proportion of oxidized Fe^{3+}; typically these oxides exist as hematite or maghemite. These are less magnetic and are more difficult to separate via physical methods, often requiring chemical extraction to separate them from the raw mineral (Svoboda, 1982).

Beneficiated and concentrated ore containing in excess of 71 wt.% iron content is typically required for iron metallurgy (Oeters *et al.*, 2000). Iron oxides obtained via magnetic separation typically range from 40 to 60 wt.% of Fe_2O_3, corresponding to an iron content of approximately 28–42 wt.%. This is well short of the required quality of iron ore raw material for iron production, and has to be further upgraded for it to be useful. Chemical leaching of the serpentine mineral can potentially yield iron oxides with higher purity, but these methods can be very energy intensive.

---

[5]Random close packing behaviour suggests that around 64% of the volume of asphalt would be occupied by coarse aggregates if the solids were filled randomly, leaving behind around 35% volume to be filled by bitumen, fine aggregates and air void. Since air void is ideally at 4% and bitumen is only a small fraction of the asphalt, which leaves approximately 25–30% volume to be filled by fine aggregates.

World production of pig iron (an intermediate material refined from iron ore, often intended for downstream processing into wrought iron or steel) in 2014 was nearly 1.2 BT (USGS, 2015d). This corresponds to approximately 2–2.3 BT of iron ore processed, comfortably more than that which can be produced from MC. Therefore, it can be concluded that the iron oxide obtained from serpentine minerals can be integrated (without significant disruption) into the worldwide iron ore market.

Depending on local conditions, the product mixture can expect to sell at roughly $10 per tonne as a construction aggregate to compete with readily available sand products (USGS, 2015d). The price of iron ore has been steadily declining since 2014, due to a supply glut in China and Australia (Haque *et al.*, 2015).

5.4.2 *Fine Applications of MC Products*

Products from MC can command higher prices if they are intended for fine applications. However, they have to be of a higher quality and purity than those intended for bulk applications. If a MC process is capable of producing products with sufficiently high quality, these products can generate a considerable amount of revenue to sustain wider MC operations.

Calcium and magnesium carbonate produced from aqueous pH-swing MC processes can be exceptionally pure, often capable of exceeding 99% purity (Hemmati *et al.*, 2014). These carbonate products can easily replace conventionally produced materials in some of the high-end applications such as paper coatings. This subsection discusses the uses of *highly pure materials* refined from the components present in serpentine. These materials include calcium and magnesium carbonate, refined silica (typically precipitated), and ferric oxide.

5.4.2.1 *Alkaline-Earth Carbonates*

Calcium carbonate and magnesium carbonate share many similar properties, thus there is a significant overlap in their respective downstream applications. Both calcium and magnesium carbonate find their largest use in the manufacture of lime (CaO) and magnesia (MgO). Unfortunately, this means that they are calcined to release any bound CO_2, thus immediately ruling this out as a potential disposal pathway.

However, these carbonates are often used as filler and brightening agents for coated paper, and can constitute up to 25% of the total paper weight (Carr *et al.*, 2000). In 2013, the reported consumption of calcium carbonate in Europe for the manufacturing of coated papers was around 8.8 Mt (CEPI, 2013). This would be equivalent to about 4.6 million tonnes of CO_2 utilised for coated paper production in Europe.

Calcium carbonate is also used as a raw material to produce soda ash (sodium carbonate) via the Solvay process (Thieme, 2000). In the Solvay process, sodium chloride and calcium carbonate undergo metathesis to produce sodium carbonate and calcium chloride. The calcium chloride is often treated as a waste product to be disposed of, but it can potentially be used as a source of brine to further sequester CO_2.

$$2NaCl + CaCO_3 \rightarrow Na_2CO_3 + CaCl_2 \qquad (5.3)$$

World soda ash production in 2014 was 51.6 Mt, of which nearly 72% (37 Mt) were synthesised mainly via the Solvay process (USGS, 2015e). Based on the mass balance, the Solvay process consumes around 35 Mt of calcium carbonate, which corresponds to an equivalent of 15.3 Mt of CO_2 captured and utilised.

An emerging use for calcium carbonate is the production of calcite boards for construction purposes. Calera has recently developed a process to synthesize vaterite (a semi-stable form of calcium carbonate), using steel slag and other industrial wastes as the calcium source (Monteiro *et al.*, 2013). In their process, calcium is extracted from the slag and then reacts with CO_2 scrubbed from flue gas to produce vaterite, which is then cured to the more stable calcite. It might be possible for these calcite products to act as substitutes for plywood or in other non-load bearing structural applications.

To a lesser degree, these carbonates can also be incorporated into plastics to improve various properties of the plastic product. The addition of alkaline-earth carbonates to plastics imparts a degree of flame resistance and smoke inhibition to the product. In the case of PVC, calcium carbonate also acts as a filler material, constituting up to 60% of the plastic weight depending on the formulation. Addition of calcium carbonate changes the physical properties of the plastic product, improving its colour, strength and rigidity. By far, the most common type of filler, several million tonnes of calcium carbonate are used in the manufacture of plastics every year.

High-purity calcium and magnesium carbonates can be expected to be sold at higher prices than the carbonate/silica mixture, typically around a few hundred US dollars per tonne material depending on quality.

5.4.2.2 *Refined Silica*

The coarse silica residue after acid leaching in certain pH-swing MC processes can be further refined into high grade, precipitated silica. Precipitated silica is probably the easiest type of refined silica product to manufacture from coarse silica, seeing as the other types of silica products such as fumed silica or aerogels are often much more energy intensive and technically complex to manufacture. In certain pH-swing processes where the serpentine mineral is leached with acid, the solid residue is usually fairly concentrated in SiO_2 (>80 wt.%), making it easy to extract and refine (Bai *et al.*, 2014). Precipitated silica is widely used in a variety of industries, mainly as additives and filler in rubber and polymers, as well as in other special formulations for food, personal care products and pharmaceuticals. One of the largest consumers of precipitated silica is the tire industry, where it is used as filler or additive to enhance the properties of the rubber (Flörke *et al.*, 2000).

A Best Available Techniques (BAT) reference document, prepared by the Institute for Prospective Technological Studies (IPTS), reports a production capacity of 620,000 tonnes of synthetic amorphous silica per year in the EU, of which around 73% is in the form of precipitated silica (the others being pyrogenic silica or silica gel) (IPTS, 2007). In other words, this means that the EU produces up to 450,000 tonnes of precipitated silica per year. The IPTS also reports that the EU produces around 30% of the world's silica; assuming the production breakdown of silica types is similar across different regions, it thus follows that the worldwide demand for precipitated silica would be roughly 1.5 Mt per year.

Ceramics, silica gels and refractories are much smaller (but nevertheless important) markets for precipitated silica. These applications usually have market sizes on the order of thousands of tonnes per year. Though the coarse silica can perhaps be sold at a price of around \$10 per tonne as raw material for glassmaking, a pre-purified silica material can fetch much higher prices (easily going up to several hundred dollars a tonne of precipitated silica), significantly adding value to any MC process that can co-produce it at low cost.

5.4.2.3 *Iron and Other Metals*

Iron in serpentine can be extracted as an oxide or hydroxide, depending on the specific process technology for MC. Production of high grade iron compounds typically require chemical leaching of the serpentine mineral and precipitation from solution. The most important applications of pure iron oxides and hydroxides include use as pigments, as a component of thermite and production of magnetic materials.

Iron oxides and hydroxides are widely used because they are non-toxic and do not degrade easily over time. Ullmann's Encyclopedia of Industrial Chemistry gives an excellent overview of iron-based pigments, under its inorganic pigments chapter (Völz, 2000). Various forms of iron oxides and hydroxides are used as pigments, with specific compositions and crystal structures. These are sometimes blended with other metal oxides to obtain the desired colours. Some of the most common iron oxides and hydroxides that are used as pigments are: goethite (α-FeOOH), lepidocrocite (γ-FeOOH), hematite (α-Fe$_2$O$_3$), maghemite (γ-Fe$_2$O$_3$) and magnetite (Fe$_3$O$_4$) (Swiler, 2000).

The colours of iron-based pigments usually tend towards earthy hues, ranging from yellow to brown or red. While these iron-based pigments can be found naturally and are mined in various locales, they are often produced synthetically as the manufacturing processes offer a larger degree of control over the properties of these pigments. The majority of iron-based pigments are now manufactured, using cheap raw materials such as scrap metal and pickling waste to synthesise higher value-added materials (Stephenson *et al.*, 1984).

The colours of iron-based pigments are affected by their composition, particle size, oxidation state, crystal structure and impurities present. However, small variations (1–2% difference) in these properties generally do not affect the overall quality of the pigment significantly. Most of these iron-based pigments find use in construction materials, paint and coatings, with a lesser degree being incorporated into plastics and rubber materials. Several hundred kilotonnes of iron oxides are produced annually for use as pigments.

Apart from pigments, iron oxide (more specifically magnetite and maghemite) is also used as magnetic nanoparticles for various applications. Among the more common uses for iron oxide nanoparticles include tag agents for magnetic resonance imaging (MRI) and manufacture of magnetic storage devices (Gupta and Gupta, 2005). Obviously these applications represent only a miniscule fraction of the total demand for iron oxides, but the stringent quality specifications required for the applications mean that the price for these materials are often very high.

Iron oxide can also be used as a cheap oxidising agent in thermite. Thermite is mainly used for high temperature welding of steel, and also for pyrometallurgical refining of various ores. Both hematite and magnetite can be mixed with aluminium powder to produce thermite.

$$Fe_2O_3 + 2Al \rightarrow Al_2O_3 + 2Fe + Heat$$

Since making thermite simply involves mixing together two powders (one reactive metal and one less reactive oxide), it is not manufactured and stocked. Thermite is also very dangerous if handled incorrectly, thus its components are often kept separately for safety reasons. As such, not much information is available regarding iron oxide use in thermite applications. However, this amount is not expected to be larger than several hundred tonnes per year.

In addition to iron, other elements are also present in varying quantities in serpentine minerals. The most abundant of these elements include aluminium, nickel, titanium and chromium, which also occur as oxides in the mineral. These metal oxides can be extracted with acid in a manner similar to magnesium and iron in the mineral. These metals may then be recovered as a salt via ion exchange or liquid–liquid extraction or oxide via neutralisation and precipitation.

The recovered metal salts are usually used in downstream electrowinning or thermal reduction processes, where it is converted into the metal. If recovered as metal oxides, these materials can have desirable physical properties and often exhibit a dazzling array of colours depending on the species. As such, these metal oxides are often sold as pigments and abrasives. Unfortunately the recovery of these elements as salts or oxides from the leachate may prove to be difficult, as they are often co-precipitated from the solution together with the iron hydroxides. Also, the relatively low concentrations may make the recovery process uneconomical. Nevertheless, if these metals (or their salts and oxides) can be obtained at high purity and large enough volumes, they can be sold to generate a substantial revenue stream to recover MC costs.

Aluminium oxide is mainly used as an abrasive, or as raw material for refining into aluminium metal. Since aluminium oxide (and its ore, bauxite) is very common, it can be mined and produced cheaply. World reserves of bauxite are around 55–75 BT; there is no scarcity of the raw material and the price of aluminium oxide is therefore quite low (around \$370 per tonne in the US, as of 2014) (USGS, 2015f). Aluminium metal is produced mainly from raw bauxite via the Bayer process followed by electrorefining, and production of the metal in 2014 was around 65 Mt of metal (USGS, 2015g). Depending on the region, significant amounts of scrap aluminium are also recycled for new metal production.

Like aluminium oxide, titanium dioxide is also fairly abundant but to a lesser degree. World reserves of titanium-containing ore (i.e. rutile) are estimated to be at about 2 BT. Refined titanium dioxide is mainly used as a whitening agent or pigment, and more than 6.5 Mt of the material was produced worldwide in 2014. World production of titanium metal was nearly 200,000 tonnes in 2014.

Nickel oxide is mainly used as pigment, and is less important in the production of nickel metal. Sulphide ores represent a major source of nickel for the metal production process, but this is expected to be gradually overtaken by common but low grade lateritic nickel ores (Whittington and Muir, 2000). Chromium oxide, more specifically Cr_2O_3, is commonly used as a pigment or abrasive. The combined demand for these two oxides is estimated to be less than 50,000 tonnes per year (not including uses intended for metal production). Discounting their use as oxides, nickel and chromium find the largest use in the manufacturing of stainless steel and other specialty alloys.

The 300 Series (or austenitic series) make up nearly two-thirds of the global production for stainless steel (Gerdemann *et al.*, 2007; ISSF, 2015). Of these, the most commonly produced are the SS304 and SS316 grades, and the typical nickel and chromium content in these stainless steel grades is 8 and 18%, respectively. Assuming that these two grades make up half of the total market demand for stainless steel (around 42 million tonnes in 2014), the total amount of nickel and chromium used in these products would be approximately 1.7 and 3.8 million tonnes, respectively. This represents a considerable amount of demand for these metals.

To conclude the discussion on market sizes and applications for the products of MC, the reader is urged to keep in mind the context under which these numbers are discussed. These numbers are mainly intended to be compared with the overall scale of MC, which we have earlier determined to be (conservatively) sized at approximately 3.5 Gt of CO_2 sequestered per year, giving around 17.5 Gt of products as a result of the various processes involved. These products can be of varying quality and quantities, depending on the type of MC process applied to achieve CO_2 sequestration. It has been shown that, in theory, there is no shortage in demand for the carbonate/silica mixture in various earthwork projects. The structures built from these earthwork projects include levees, dams, seawalls and even new land, therefore indirectly making long-term contributions towards adaptation measures to mitigate economic losses from climate change.

Sale of higher quality products from MC can also generate a sizeable income stream in the shorter term, and the revenue obtained from sale of these products can greatly improve the economics of MC and facilitate the wider adoption of this technology. In the discussions relating to the market sizes of various side-products of MC, several industries that consume these products immediately stand out from the rest. These are the soda ash, paper, tire, pigment and stainless steel manufacturing industries. They represent the largest consumers of the alkaline-earth carbonate, precipitated silica and metal oxide side-products from MC, and have worldwide capacities in excess of several million tonnes per year.

5.5 Versatility of MC Processes

Compared with other traditional chemical engineering processes, mineral carbonation is still in its infancy and can be considered as rather underdeveloped. The big drawback to this is that there are no large scale commercial MC operations, and only a few working prototypes or demonstration plants. However, this also means that there is no commonly accepted "best" process that dominates the field, in the manner of the Bayer process for alumina, the Kraft process for pulp, or the Solvay process for soda ash production. This allows for some impressively creative ideas to emerge from different research teams in the field, increasing the potential for innovative breakthroughs and dramatic improvement.

Perhaps, the most well-researched and iconic MC process to date is the NETL pressure carbonation process (Gerdemann *et al.*, 2007). It has been adopted by some companies for further development, and is the benchmark for improvement studies by many other researchers. Other conceptually different processes, most notably the aqueous pH-swing processes, have also emerged after it and are currently in various stages of development. These two schools of thought in mineral carbonation seek to achieve the same purpose: to permanently sequester CO_2 as a mineral carbonate, but their approaches and outcomes are very different.

Pressure carbonation processes typically have relatively low CO_2 penalties due to their simplicity, resulting in higher CO_2 sequestration efficiencies. However, these processes tend to be uneconomical because their products are of low quality, and cannot be sold to recover the processing costs. On the other hand, aqueous pH-swing processes are much more energy intensive and complex in design, and often have lower CO_2 sequestration efficiencies due to the embodied emissions of the chemicals used in operations. This however allows for production of higher quality products, which can be sold for a large profit.

These two types of processes stand in stark contrast in terms of economics and efficiency, though each is quite successful at what they are designed for. However, at present, they remain incapable of achieving a fair balance between the economics and efficiency within the process. In other words, this means that specialisation in these individual process designs have made other aspects of the process much more difficult to improve, at least without sacrificing some of the advantage that makes it attractive at the same time. Attempts to improve the economics of the process often involve compromising and reducing the overall efficiency (and *vice versa*), and the sacrifice in one aspect usually does not justify the gain in limited benefits in the other.

One of the solutions to this dilemma is an integration or hybridisation of these two types of processes, such that the strengths of one process are introduced and exploited to cover for the weakness of the other.

Figure 5.3 depicts a symbiotic relationship between these two conceptually different types of processes. Here, they co-exist as a larger, more holistic mineral carbonation process, where the bulk of the sequestration takes place in the pressure carbonation segment and a small fraction of the minerals is diverted into a smaller aqueous pH-swing side process. The amount of CO_2 that is sequestered in the pressure carbonation segment has to be sufficiently large, such that

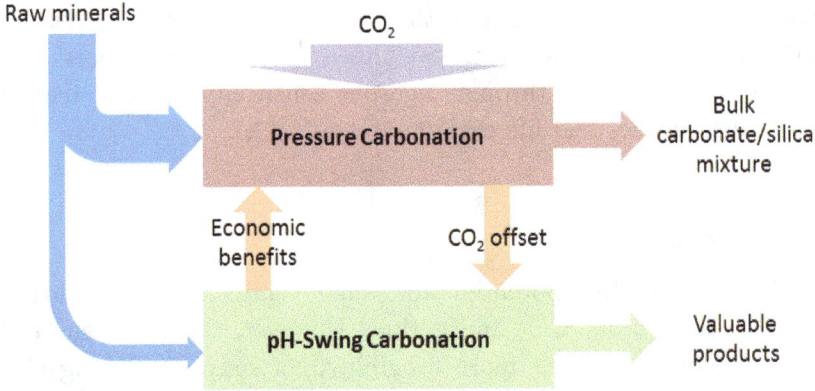

Figure 5.3 Cooperation between processes for maximum environmental and economic benefits.

it is capable of offsetting any CO_2 penalties that are incurred by the pH-swing side process. The role of the smaller side process is to generate a large stream of revenue for the purposes of recovering the costs associated with the overall integrated carbonation process.

The benefits in this arrangement are evident when compared with the case where each segment of the integrated process is conducted in isolation. Despite the promising technical results, pressure carbonation on its own would struggle to obtain sufficient private sector funding for widespread implementation. It would perhaps be unable to even attract enough interest for the construction of a pilot or demonstration plant, unless the associated costs are dramatically reduced. On the other hand, pH-swing carbonation would probably be attractive enough to see some small scale plants being built in the near future, but its limitations in terms of CO_2 sequestration capacity would quickly be apparent. Furthermore, at the requisite scales for meaningful CO_2 reductions, its products would flood the global markets, leading to price reductions to the extent that the revenue from these products are no longer sufficient to sustain long-term operations for pH-swing carbonation.

In other words, these two types of processes are unlikely to succeed on their own. However, when combined, they can form a synergistic partnership where they can compensate for each other's shortcomings. In order to give the best possible economic benefits and sequestration efficiency, the weightage given to each process type has to be determined through rigorous sensitivity studies, techno-economic analyses and life cycle analyses (LCA).

5.6 Conclusions

MC has incredible potential to make significant contributions towards large scale CO_2 mitigation. However, for it to succeed, it must offer something more tangible than simply reducing CO_2 emissions. MC has several advantages compared with other proposed sequestration options, these are: (1) permanence; (2) scalability and (3) versatility. By leveraging these strengths, MC can become a leading option to sequester the majority of CO_2 currently being emitted.

Minerals suitable for MC are widely distributed around the world, and CO_2 stored as inorganic carbonates can be considered to be permanently removed from the atmosphere, since the carbonate products are very stable under a wide range of conditions.

MC is a scalable technology; it is capable of matching the CO_2 sequestration volumes required. The required operations are compa-rable with some already-existing industries, such as the construction, mining and iron and steel manufacturing. The products of MC are versatile, and depending on the specific process employed, the products can either occur as a bulk carbonate/silica mixture or highly pure materials for use in other industrial processes. The uses for the bulk carbonate/silica mixture include earthwork projects (levees, dams, land reclamation) and road pavement. Iron oxide recovered from the mineral can be sold as raw material for iron production. Unfortunately, despite being a good match scale-wise, the carbonate/silica mixture is unlikely to be suitable for use as concrete aggregate due to its chemical and physical properties.

With aqueous pH-swing processes, higher quality materials are obtainable, and these can fetch higher prices. However, their market sizes are much smaller, and these can easily be flooded by the output of mineral carbonation. Nevertheless, some of the potential markets for high quality products from MC are the soda ash, paper, tire, pigment and stainless steel manufacturing industries. These are the larger industries that consume several million tonnes' worth of raw materials per year.

Stand-alone MC processes are unlikely to succeed on their own. Combinations of pressure carbonation and pH-swing carbonation processes are suggested, such that they co-exist within an integrated process. In this integrated process, the pressure carbonation segment takes on the main burden of the sequestering CO_2, while the pH-swing carbonation segment is designed to maximise the revenue for the overall process. Together, they can form a partnership where the strengths of each process are leveraged to benefit each other.

References

Abts, G., T. Eckel and R. Wehrmann (2000). Polycarbonates. In: *Encyclopedia of Industrial Chemistry* [Ullmann, (ed.)]. Wiley-VCH Verlag GmbH & Co. KGaA.

Anderson, K. and A. Bows (2011). Beyond 'dangerous' climate change: Emission scenarios for a new world. *Philosophical Transactions of the Royal Society of London A: Mathematical, Physical and Engineering Sciences* 369(1934): 20–44.

ASTM (2004). *Standard Specification for Concrete Aggregates*. ASTM International West Conshohocken, PA.

Bai, P., P. Sharratt, T. Y. Yeo and J. Bu (2014). A facile route to preparation of high purity nanoporous silica from acid-leached residue of serpentine. *Journal of nanoscience and nanotechnology* 14(9): 6915–6922.

Brady, K. U., A. R. Kruckeberg and H. Bradshaw Jr (2005). Evolutionary ecology of plant adaptation to serpentine soils. *Annual Review of Ecology, Evolution, and Systematics* 36: 243–266.

Caillaud, J., D. Proust and D. Righi (2006). Weathering sequences of rock-forming minerals in a serpentinite: influence of microsystems on clay mineralogy. *Clays and Clay Minerals* 54(1): 87–100.

Carr, F. P., D. K. Frederick and U. B. Staff (2000). Calcium Carbonate. In: *Encyclopedia of Chemical Technology* [Kirk-Othmer (eds.)]. John Wiley & Sons, Inc.

CEPI (2013). Key Statistics European Pulp and Paper Industry. Available from: http://www.cepi.org/system/files/public/documents/publications/ statistics/ 2014/Final%20Key%20statistics%202013.pdf

Chen, Y., J. P. Syvitski, S. Gao, I. Overeem and A. J. Kettner (2012). Socio-economic impacts on flooding: A 4000-year history of the Yellow River, China. *Ambio* 41(7): 682–698.

Colbert, B. and Z. You (2012). The determination of mechanical performance of laboratory produced hot mix asphalt mixtures using controlled RAP and virgin aggregate size fractions. *Construction and Building Materials* 26(1): 655–662.

Flörke, O. W., H. A. Graetsch, F. Brunk, L. Benda, S. Paschen, H. E. Bergna, W. O. Roberts, W. A. Welsh, C. Libanati, M. Ettlinger, *et al.* (2000). Silica, in: (Eds.), Ullmann's *Encyclopedia of Industrial Chemistry.* Wiley-VCH Verlag GmbH & Co. KGaA.

Gerdemann, S. J., W. K. O'Connor, D. C. Dahlin, L. R. Penner and H. Rush (2007). *Ex situ* aqueous mineral carbonation. *Environmental science & technology* 41(7): 2587–2593.

Glass, G. and N. Buenfeld (1997). The presentation of the chloride threshold level for corrosion of steel in concrete. *Corrosion Science* 39(5): 1001–1013.

Gupta, A. K. and M. Gupta (2005). Synthesis and surface engineering of iron oxide nanoparticles for biomedical applications. *Biomaterials* 26(18): 3995–4021.

Haque, M. A., E. Topal and E. Lilford (2015). Iron ore prices and the value of the Australian dollar. *Mining Technology* 1743286315Y: 0000000008.

Hemmati, A., J. Shayegan, J. Bu, T. Y. Yeo and P. Sharratt (2014). Process optimization for mineral carbonation in aqueous phase. *International Journal of Mineral Processing* 130: 20–27.

Hemmati, A., J. Shayegan, P. Sharratt, T. Y. Yeo and J. Bu (2014). Solid products characterization in a multi-step mineralization process. *Chemical Engineering Journal* 252: 210–219.

Hendriks, C. and K. Blok (1993). Underground storage of carbon dioxide. *Energy Conversion and Management* 34(9): 949–957.

IPTS (2007). Large Volume Inorganic Chemicals — Solids and Others Industry: Best Available Techniques (BAT) reference document.

ISSF (2015). ISSF — Stainless Steel in Figures (2015).

Jaramillo, P., W. M. Griffin and S. T. McCoy (2009). Life cycle inventory of CO₂ in an enhanced oil recovery system. *Environmental Science & Technology* 43(21): 8027–8032.

Kelemen, P. B. and J. Matter (2008). *In situ* carbonation of peridotite for CO₂ storage. *Proceedings of the National Academy of Sciences* 105(45): 17295–17300.

Kosmatka, S. H. and M. L. Wilson (2011). Design and Control of Concrete Mixtures: The guide to applications, methods, and materials. p. 458, ISBN: 893122726. Available from: https://trid.trb.org/view.aspx?id=1120874

Leaton, J. (2012). Unburnable Carbon — Are the world's financial markets carrying a carbon bubble. Available from: https://www.carbontracker.org/wp-content/uploads/2014/09/Unburnable-Carbon-Full-rev2-1.pdf. Access 20 May 2016.

Levina, E., S. Bennett and S. McCoy (2013). IEA Technology roadmap: Carbon capture and storage. Available from: http://www.iea.org/publications/free publications/publication/technology-roadmap-carbon-capture-and-storage-2013.html. Access 5 April 2016.

Liu, X.-M., G. Lu, Z.-F. Yan and J. Beltramini (2003). Recent advances in catalysts for methanol synthesis via hydrogenation of CO and CO_2. *Industrial & Engineering Chemistry Research* 42(25): 6518–6530.

Mavrovic, I., A. R. Shirley and G. R. B. Coleman (2000). Urea. In: *Encyclopedia of Chemical Technology* [Kirk-Othmer (eds.)]. John Wiley & Sons, Inc.

Meessen, J. H. (2000). Urea. In: *Encyclopedia of Industrial Chemistry* [Ullmann (ed.)]. Wiley-VCH Verlag GmbH & Co. KGaA.

Miliutenko, S., A. Björklund and A. Carlsson (2013). Opportunities for environmentally improved asphalt recycling: the example of Sweden. *Journal of Cleaner Production* 43: 156–165.

Monteiro, P. J., L. Clodic, F. Battocchio, W. Kanitpanyacharoen, S. R. Chae, J. Ha and H.-R. Wenk (2013). Incorporating carbon sequestration materials in civil infrastructure: A micro and nano-structural analysis. *Cement and Concrete Composites* 40: 14–20.

Najafabadi, A. T. (2013). CO_2 chemical conversion to useful products: an engineering insight to the latest advances toward sustainability. *International Journal of Energy Research* 37(6): 485–499.

National Asphalt Paving Association (2011). *The Asphalt Paving Industry: A Global Perspective*. EAPA, Brussels.

Oeters, F., M. Ottow, D. Senk, A. Beyzavi, J. Güntner, H. B. Lüngen, M. Koltermann and A. Buhr (2000). Iron, 1. Fundamentals and principles of reduction processes. In: *Encyclopedia of Industrial Chemistry* [Ullmann (ed.)]. Wiley-VCH Verlag GmbH & Co. KGaA.

Pachauri, R. K., M. Allen, V. Barros, J. Broome, W. Cramer, R. Christ, J. Church, L. Clarke, Q. Dahe and P. Dasgupta (2014). *Climate Change 2014: Synthesis Report*. Contribution of Working Groups I, II and III to the Fifth Assessment Report of the Intergovernmental Panel on Climate Change.

PCA (2015). How Concrete Is Made. Retrieved 4/7/2015. Available from: http://www.cement.org/cement-concrete-basics/how-concrete-is-made.

Railsback, B. (2015). Some Fundamentals of Mineralogy and Geochemistry. Retrieved 7/4/2015. Available from: http://www.gly.uga.edu/railsback/FundamentalsIndex.html.

Sanna, A., M. Dri and M. Maroto-Valer (2013). Carbon dioxide capture and storage by pH swing aqueous mineralisation using a mixture of ammonium salts and antigorite source. *Fuel* 114: 153–161.

Sanna, A., M. R. Hall and M. Maroto-Valer (2012). Post-processing pathways in carbon capture and storage by mineral carbonation (CCSM) towards the introduction of carbon neutral materials. *Energy & Environmental Science* 5(7): 7781–7796.

Sanna, A., M. Uibu, G. Caramanna, R. Kuusik and M. Maroto-Valer (2014). A review of mineral carbonation technologies to sequester CO_2. *Chemical Society Reviews* 43(23): 8049–8080.

Shand, M. A. (2006). *The Chemistry and Technology of Magnesia*. John Wiley & Sons.

Singstat (2014). Statistics Singapore. Retrieved 4/7/2015. Available from: http://www.singstat.gov.sg/statistics/latest-data#14.

Speight, J. G. (2000). Asphalt. In: *Encyclopedia of Chemical Technology* [Kirk-Othmer (eds.)]. John Wiley & Sons, Inc.

Sprung, S. and J. Kropp (2000). Cement and Concrete. In: *Encyclopedia of Industrial Chemistry* [Ullmann (ed.)]. Wiley-VCH Verlag GmbH & Co. KGaA.

Stephenson, J., J. Hogan and R. Kaplan (1984). Recycling and metal recovery technology for stainless steel pickling liquors. *Environmental Progress* 3(1): 50–53.

Stern, N. H. (2006). *Stern Review: The Economics of Climate Change.* HM treasury, London.

Streit, J. E. and R. R. Hillis (2004). Estimating fault stability and sustainable fluid pressures for underground storage of CO_2 in porous rock. *Energy* 29(9–10): 1445–1456.

Svoboda, J. (1982). Magnetic flocculation and treatment of fine weakly magnetic minerals. *Magnetics, IEEE Transactions* 18(2): 796–801.

Swenson, E. G. and J. E. Gillott (1964). Alkali-carbonate rock reaction. *Highway Research Record* 45: 21–40.

Swiler, D. R. (2000). Pigments, Inorganic. In: *Encyclopedia of Chemical Technology* [Kirk-Othmer (eds.)]. John Wiley & Sons, Inc.

Tao, Z. and A. Clarens (2013). Estimating the Carbon sequestration capacity of shale formations using methane production rates. *Environmental Science & Technology* 47(19): 11318–11325.

Thieme, C. (2000). Sodium Carbonates. In: *Encyclopedia of Industrial Chemistry* [Ullmann (ed.)]. Wiley-VCH Verlag GmbH & Co. KGaA.

Tkalich, P., P. Vethamony, Q.-H. Luu and M. Babu (2013). Sea level trend and variability in the Singapore Strait. *Ocean Science* 9: 293–300.

Uehara, S. and H. Shirozu (1985). Variations in chemical composition and structural properties of antigorites. *Mineralogical Journal* 12(7): 299–318.

US Army, C. O. E. (2015). Levee Systems. Retrived 4/7/2015. Available from: http:// www.mvd.usace.army.mil/About/MississippiRiverCommission%28 MRC%29/MississippiRiverTributariesProject%28MRT%29/LeveeSystems. aspx.

USGS (2015a). USGS — Minerals Information: Cement. Retrieved 4/7/2015. Available from: http://minerals.usgs.gov/minerals/pubs/commodity/cement/index.html.

USGS (2015b). USGS – Minerals Information: Crushed Stone. Retrieved 4/7/2015. Available from: http://minerals.usgs.gov/minerals/pubs/commodity/stone_crushed/mcs-2015-stonc.pdf.

USGS (2015c). USGS — Minerals Information: Iron and Steel. Retrieved 4/7/2015. Available from: http://minerals.usgs.gov/minerals/pubs/commodity/iron_&_steel/mcs-2015-feste.pdf.

USGS (2015d). USGS — Minerals Information: Iron Ore. Retrieved 4/7/2015. Available from: http://minerals.usgs.gov/minerals/pubs/commodity/iron_ore/mcs-2015-feore.pdf.

USGS (2015e). USGS — Minerals Information: Soda Ash. Retrieved 4/7/2015. Available from: http://minerals.usgs.gov/minerals/pubs/commodity/soda_ash/mcs-2015-sodaa.pdf.

USGS (2015f). USGS — Minerals Information: Aluminum. Retrieved 4/7/2015. Available from: http://minerals.usgs.gov/minerals/pubs/commodity/alumi num/mcs-2015-alumi.pdf.

USGS (2015g). USGS — Minerals Information: Bauxite and Alumina. Retrieved 4/7/2015. Available from: http://minerals.usgs.gov/minerals/pubs/commodity/bauxite/mcs-2015-bauxi.pdf.

Völz, H. G. (2000). Pigments, Inorganic, 1. General. In: *Encyclopedia of Industrial Chemistry* [Ullmann (ed.)]. Wiley-VCH Verlag GmbH & Co. KGaA.

Whittington, B. and D. Muir (2000). Pressure acid leaching of nickel laterites: A review. *Mineral Processing and Extractive Metullargy Review* 21(6): 527–599.

Wood, S. and A. Cowie (2004). A review of greenhouse gas emission factors for fertiliser production. Research and Development Division, State Forests of New South Wales. Cooperative Research Centre for Greenhouse Accounting. For IEA Bioenergy Task 38. Available from: http://www.task38.org/publications/task38_description_2013.pdf

Index

product applications, viii
production, 158
production of sellable products, 10
products, 160
proton–metal exchange, 31
PVC, 152
pyroxene, 30

R

rate of carbonation, 95
rate-limiting mechanism, 94
raw material for production of other
    chemicals, 138
RCO_2, 6
reaction kinetics, 58
reaction of CO_2 with metal oxide, 2
reactivity difference, 31
reactivity of minerals and rocks, 28
reactivity of the minerals, 33
reclaiming land from coastal regions,
    145
reclamation, 146
reducing the atmospheric CO_2, 4
refined silica, 150, 152
refractories, 153
refractory material, 143
remediated, 119
removal of iron impurities, 56
requirements for MC, 67
resources, 17
resources available, 12
resources for MC, 5
RM is a caustic by-product, 118
rock material, 70
rotary kiln, 65

S

scalability, 135, 160
secondary benefits, 91
separate CO_2, 42
separation of CO_2 from natural gas,
    41
separation of the alkaline-earth
    carbonates, 143
separation techniques, 142

sequester CO_2 in stable insoluble
    carbonates, 36
sequestrated CO_2, 10
sequestration of CO_2, 137
serpentine(s), 3, 20, 31, 115, 144, 149,
    153
serpentine mineral, 141
serpentine structure, 32
serpentinites, 26
side-products, 157
silica, 142
Singapore, 146
Singapore's land reclamation
    projects, 145
Skyonic process, vi
Slag2PCC, 102
slow kinetics, 51
SO_2, 47
soda ash, 157, 161
sodium bisulphate solution, 59
soils and land use systems, 27
solid products, 11
solid wastes, 91
solubility of ammonium and sulphate
    in water, 76
Solvay process, 151, 157
source material, 20
stabilisation of the contaminants, 110
stainless steel, 156–157
steelmaking industry, 95
steelmaking slags, 57
stepwise approach of the gas–solid
    MC, 60
store or use CO_2, 136
sub-oceanic mantle, 23
supercritical CO_2, 137
surface area, 27, 49
symbiotic relationship, 158

T

talc, 20
temperature optimisation, 51
thermal activation, 32
thermal activation mechanisms, 30
thermal decomposition, 148

thermal pre-activation of rock
material, 54
thermal reduction processes, 155
thermite, 154
thermodynamic limitations, 47
thin film carbonation, 105
three-stage dry/wet/dry, 61
tire, 161
titanium, 155
titanium dioxide, 155
total cost to sequestering 1 tonne
CO_2, 59
total energy requirement, 96
transforming CO_2 into a different
chemical, 136
transport, 67, 69
transportation-controlled mechanism,
94
tremolite, 20
trend, 50
type of waste, 120

U

ultramafic rocks, 19, 21–22, 24,
36, 38
unviable process, 50

urea, 138
utilisation of Mg-rich tailings, 114
utilisation of microbial medium, 28
utilisation of wastes, 102
utilisation pathways, 134
utilisation routes, 11

V

valorisation of metallurgical slag, 103
variety of pre-treatments, 8
versatility, 135, 157, 160

W

waste cement from aggregate
recycling process, 112
waste cement powder, 114
waste concrete production, 112
waste management, 121
waste streams, 7
wastes from solid fuel combustion, 93
water consumption, 76
water content, 76
water use, 76
wollastonite, 3
world soda ash production, 151

www.ingramcontent.com/pod-product-compliance
Lightning Source LLC
Chambersburg PA
CBHW050628190326
41458CB00008B/2177